Just for Openers

A Guide to Beer, Soda, & Other Openers

Donald A. Bull
& John R. Stanley

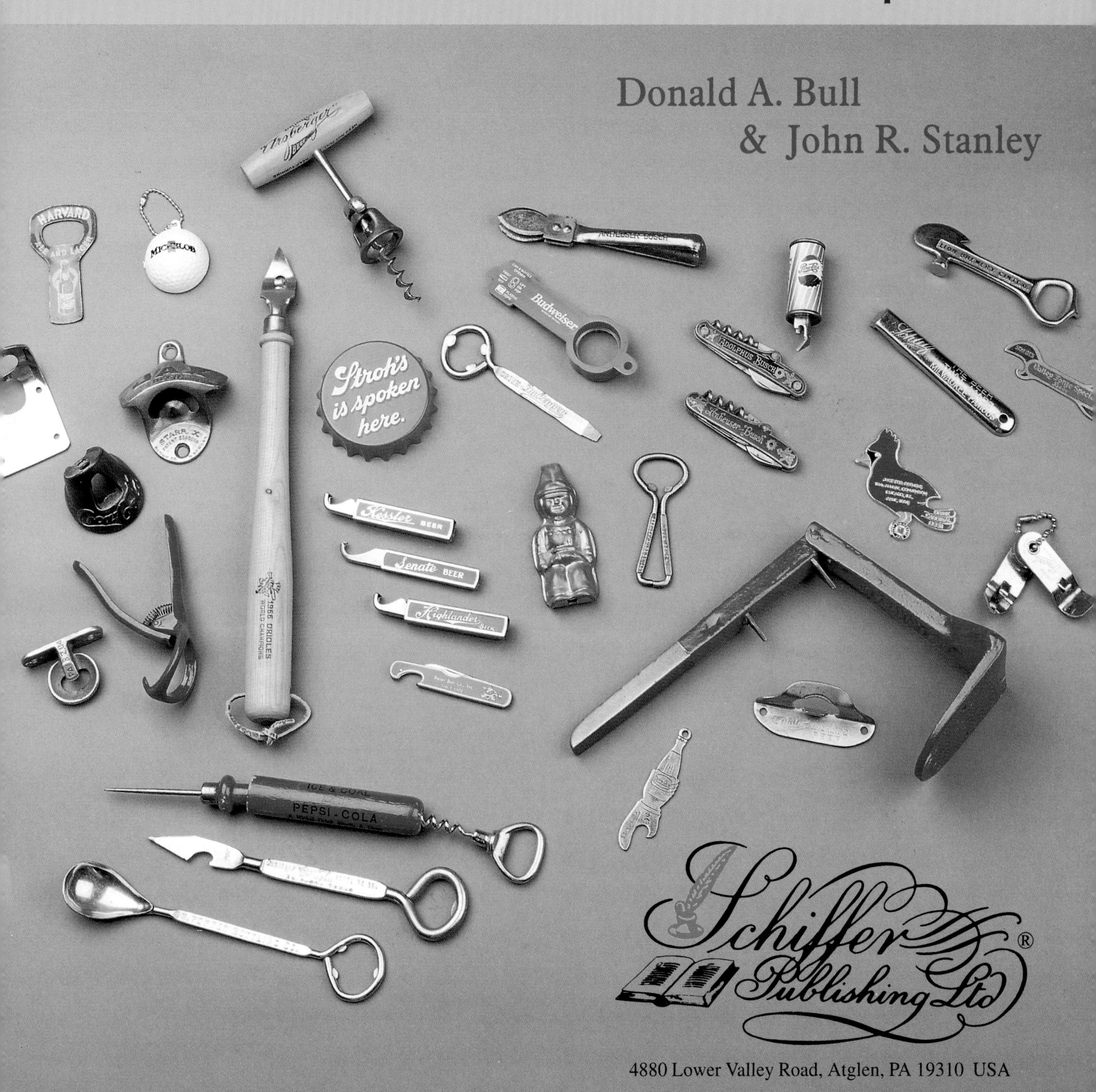

Schiffer Publishing Ltd
4880 Lower Valley Road, Atglen, PA 19310 USA

Dedication

To William Painter

CIP DATA Coming

Design by Blair Loughrey
Type set in Americana XBd BT/Aldine721 BT

ISBN: 0-7643-0846-7
Printed in China
1 2 3 4 CT

Published by Schiffer Publishing Ltd.
4880 Lower Valley Road
Atglen, PA 19310
Phone: (610) 593-1777; Fax: (610) 593-2002
E-mail: Schifferbk@aol.com
Please visit our web site catalog at
www.schifferbooks.com

This book may be purchased from the publisher.
Include $3.95 for shipping.
Please try your bookstore first.
We are interested in hearing from authors
with book ideas on related subjects.
You may write for a free catalog.

In Europe, Schiffer books are distributed by
Bushwood Books
6 Marksbury Rd.
Kew Gardens
Surrey TW9 4JF England
Phone: 44 (0)181 392-8585; Fax: 44 (0)181 392-9876
E-mail: Bushwd@aol.com

Contents

Acknowledgments

Since its formation in 1978, *Just For Openers (JFO)* has been the guiding light for beer advertising opener and corkscrew collectors. The organization is thriving and this book is a collective result of their research and efforts over the years. Gaining and sharing knowledge is the most important goal of *JFO*.

As with any endeavor of this kind, special thanks must be given to a number of individuals for their contributions. Many of the openers pictured in this book were loaned by these collectors. A special thanks to these contributors: William Arber, Mark Barren, Marc Benjamin, Larry Biehl, Linwood Blalock, John Cartwright, Gary Deachman, William Ennis, Jack Ford, Janet Goss, Fil Graff, Arthur Hanson, Ollie Hibbeler, Ben Hoffman, Ardea Horn (wife of the late John Horn), Roger Jarrell, Norm Jay, Art Johnson, Vic Keown, Joseph Knapp, Elvira McKienzie (wife of the late Bill McKienzie), Bob McNary, Larry Moter, Pete Nowicki, Jim Osborn, Bill Pattie, John Patton, Harold Queen, John Ruckstuhl, Art Santen, Don Sherman, Robert Sommer, Bob Stahly, Verne Vollrath, Don Whelan, Joe Young, and Toni Zruno-Rankin (daughter of the late Tom Zruno).

Even a formally published thank you does not express the gratitude to all those who took some of their most prized possessions out of their collections and sent them to be photographed for this book. Their displays were incomplete, in some cases up to three months, so this book could be completed. We were very touched by family members of collectors who have passed away, for they sent openers that hold immeasurable sentimental value.

May new collectors' interests be "opened" by this publication. May the openers and corkscrews pictured here someday fall into the hands of people who will appreciate the history and value of these collections. May this book teach new collectors when those who carry so much knowledge are no longer here. History will live on.

Finally, credit is given to "Corkie" the Yorkie, who spent many hours at John's side by the computer. She is the most valuable corkscrew of all.

Introduction

In the Beginning

In 1978, *Beer Advertising Openers, A Pictorial Guide* by Donald Bull was published. This book pictured over 200 types of American beer advertising openers and classified them by type and category. A catalog of known American advertising pieces by type was included in the book. In the following year, Bull founded *Just for Openers* with the introduction of a quarterly newsletter. The purpose of the newsletter was to give beer advertising opener collectors a vehicle to broaden their collections and to keep a running catalog of new finds by type and by advertising. In less than a year, membership of the organization had reached 200, a convention was held, and plans were underway for a meeting in the next year.

In 1981, Bull published *A Price Guide to Beer Advertising Openers and Corkscrews.* By then, over 400 distinct types of openers and corkscrews had been classified and thousands had been cataloged. Ed Kaye took over as Editor/Publisher of *Just for Openers* in 1984. In the fall, Kaye and Bull published *The Handbook of Beer Advertising Openers and Corkscrews*, which pictured 500 types of openers and corkscrews with a catalog listing of 9,000.

After his five years at the helm, Kaye relinquished his duty to Art Santen. In his introduction to his first newsletter, Santen wrote: "It would seem to me that everything that could be printed about openers has already appeared in *Just for Openers* (or has it)?" Santen quickly learned that was not the case and, over the next five years, he published twenty issues of *Just for Openers* filled with new discoveries, catalog additions, anecdotes, member stories, annual convention news, and opener history.

In 1994, John Stanley took on the Editor/Publisher responsibility for the very active group of opener collectors. John published several updates to *The Handbook of Beer Advertising Openers and Corkscrews.* By mid-1998, the number of distinct beer advertising opener and corkscrew types had grown to over 800.

Plotting

In the early 1990s, Bull sold his opener collection and kept his corkscrews and knives. Stanley's opener collection had grown to encompass a very large number of the openers. After a three hour trade session and a six pack of beer on a hot night in July, 1998, a plan was hatched to write a new opener/corkscrew book. The combination of their two collections and a call to collectors to supply missing types was needed to present in color all of the discoveries to date.

The plan was to include only opener and corkscrew types that advertised a brewing company, a beer brand, a malt company, or a malt brand produced in the United States. Openers advertising companies that bottled beer for breweries, beer stores, bars, and saloons would be excluded unless the type had a brewery, beer, or malt name.

A section on general advertising openers and soda openers is included in the book after the beer advertising opener types. This is a brief introduction to the various "other" openers available. Different categories are shown with a concentration on unusual types and major brands of soda such as Coca-Cola, Pepsi-Cola, Dr. Pepper, Nehi, Orange Crush, and Hires.

Within a short period of time, Stanley had gathered the collection pictured in this book and Bull had compiled the information. Several more meetings completed the task.

Results

We are now pleased to present you with the ultimate guidebook for this fascinating hobby. Within categories, we have grouped similar types together while maintaining the original assigned alphanumeric designation. In some issues of *Just for Openers*, types were depicted that did not have beer advertising or were not American brewery brands. These types are not shown and, therefore, there are several gaps in the alphanumeric system. To help locate types by alphanumeric designation, there is a type index included in the back of this book. Only the beer advertising openers in Part 1 have been indexed. It will become apparent to the reader that each chapter contains an Alpha type: Flat Figural Openers are type A; Key Shape Openers are type B; etc.

The history of the openers and corkscrews is best told through the bits of information in the chapter introductions, some facts included in captions, and the American Patent chronological list in the back of the book. Notable advertisements and highlights from some openers are included in the captions in *italics*.

About Pricing

Value ranges in this book are based on past sales both public and private, prices advertised in the media and at shows, prices realized on internet sales, estimates of current selling prices, and gut feelings. In cases where prior sale information is not readily available, value is based upon relative scarcity versus known values. We have included value ranges for each type. Please refer to the following guidelines when using the values in this book:

Condition: Values are for openers in excellent to mint condition.

Scarcity: The highest values are generally for items that are hardest to find especially in nice condition. Several types have few examples known and this is reflected in the value.

Demand: Malt companies or the more common brewery names are usually at the low end. Small breweries or breweries in business for a short time usually bring the highest value. Also certain areas of the country have more collectors looking for openers from that area, e.g., Missouri and Pennsylvania, bringing the price up. From time to time, a discovery of a hoard of openers will affect the values. For example, the value of an A-8 Knickerbocker Beer recently came under fire when a large supply (50) turned up in New York City. They were offered at $15 and the condition varied. Once this supply dries up, the price will go back up.

Neither the authors nor the publisher will be responsible for any gain or loss experienced from the use of these value guidelines.

If you would like to join *Just for Openers*, have questions about openers, and/or want to obtain copies of the catalog listings of openers, contact:

John R. Stanley
P. O. Box 64
Chapel Hill, NC 27514

If you would like further information on corkscrews, contact:

Donald A. Bull
P. O. Box 596
Wirtz, VA 24184

Part 1

Beer Advertising Openers

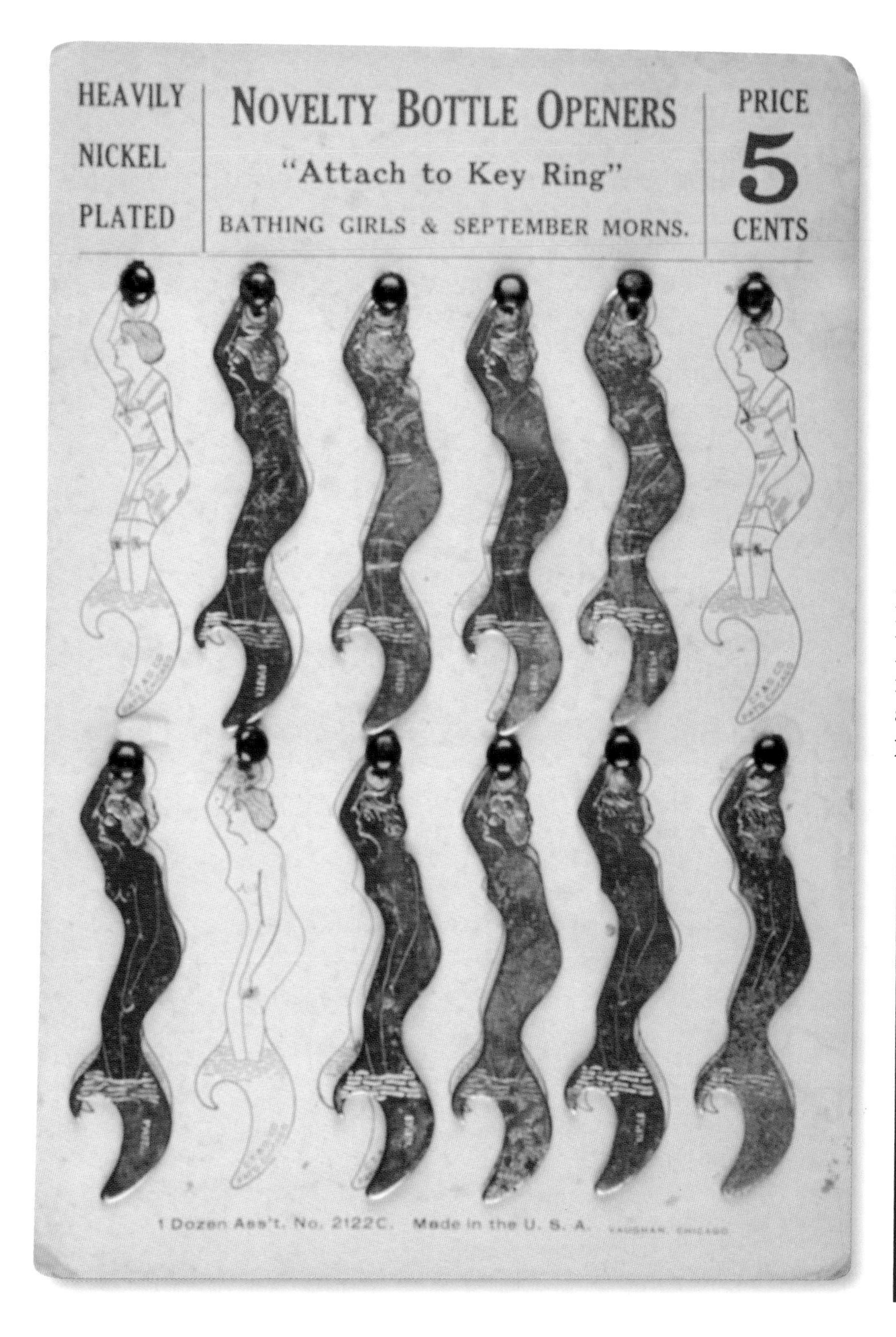

Flat Figural Openers

Openers in the shapes of bathing beauties, nudes, legs, automobiles, swords, bridges, hands, fish, various animals, and bottles were used extensively to promote brands of beers. In some cases, the designs were actually patented. Many of the shapes are found with a Prest-O-Lite key added. This is a square hole that was used to turn the valve on gas tanks of old cars for lighting the car lamps. Manufacturers include: Crown Throat and Opener Company, Chicago (later Vaughan Company); Louis F. Dow Co., St. Paul, Minnesota; J. L. Sommer Manufacturing Company, New Jersey; and Indestro Manufacturing Company, Chicago.

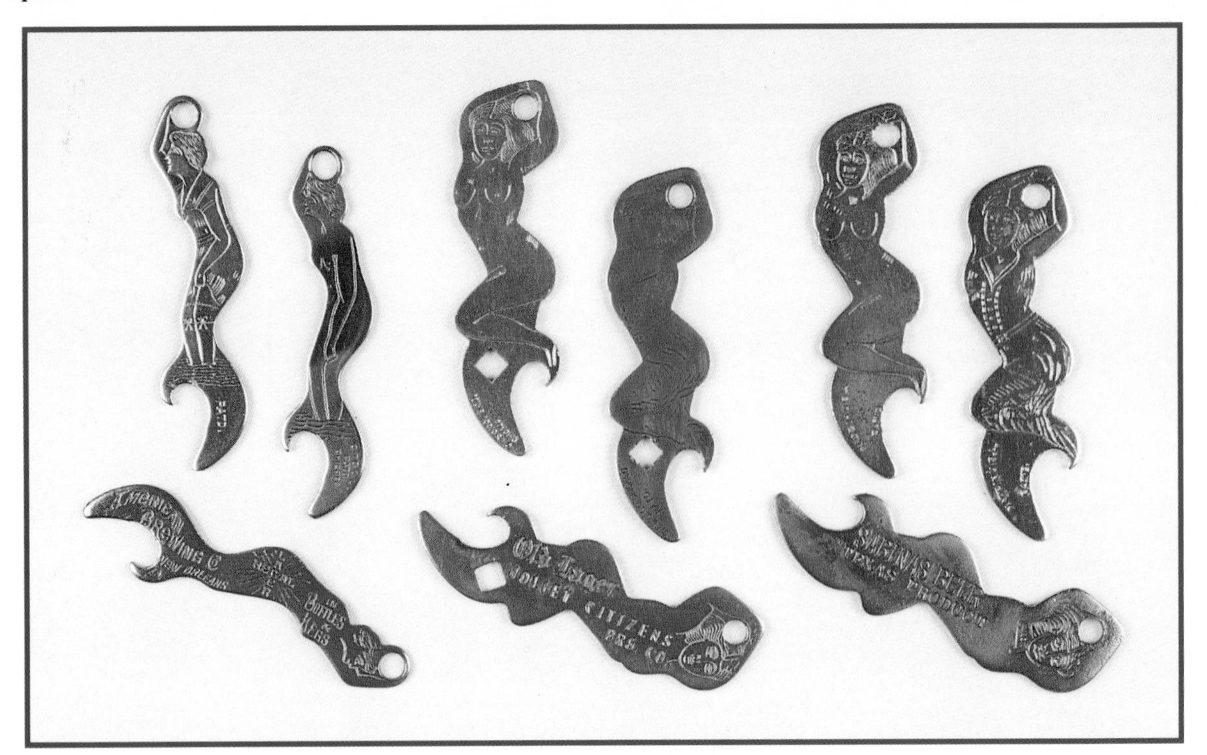

Left three: A-1 Bathing girl, mermaid, or surf-girl. Found in clothed and nude versions. Marked C. T. & O. CO. PATD. CHICAGO or PATD (C. T. & O. CO. is the abbreviation for Crown Throat & Opener Company - the first name for Vaughan Novelty). American design patent 46,762 issued to Harry L. Vaughan, December 8, 1914. 2 7/8". Prices in a 1922 Vaughan catalog were $6.50 for 250 to $15.00 per thousand in 10,000 quantity. $10-35. *The manufacturer advertised its own product on this with "Crown Throat & Opener Company, Maker, Chicago."*
Middle three: A-4 Girl clothed (calendar) and nude (Early Morn). Marked C. T. & O. CO. PAT'D. CHICAGO or PATD. American design patent 44,226 issued to Harry L. Vaughan, June 17, 1913. 2 3/4". $10-35. *The square hole is a "Prest-O-Lite" key for opening the valve on gas lamps on the running boards of old cars. One example reads "'U-Neek' Bottle Opener & Auto Gas Tank Key."*
Right three: A-5 Girl in clothed (calendar) and nude (Early Morn) versions. Marked C. T. & O. CO. PAT'D. CHICAGO or MADE IN U.S.A. PAT'D. American design patent 44,226 issued to Harry L. Vaughan, June 17, 1913. 2 3/4". $5-25.

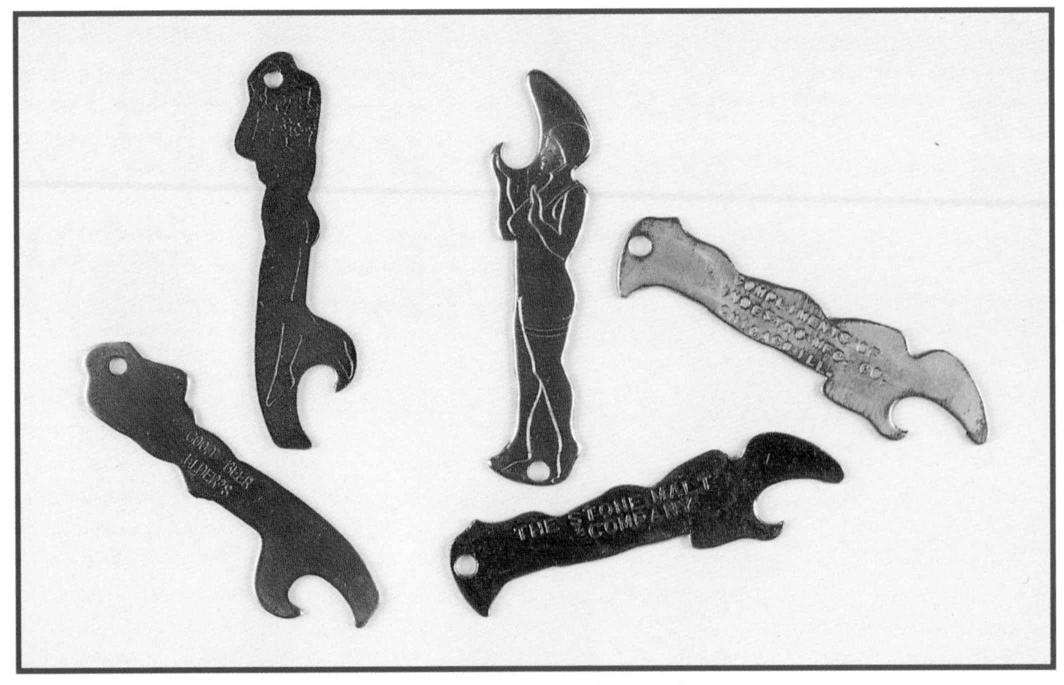

Left pair: A-2 Nude girl pouring liquid from a bottle into a glass. 3". $5-10.
Right three: A-3 Girl in bathing suit and cap. 3 1/8". $10-25. *This type is found frequently with malt company advertising. Produced by Indestro Manufacturing Company of Chicago.*

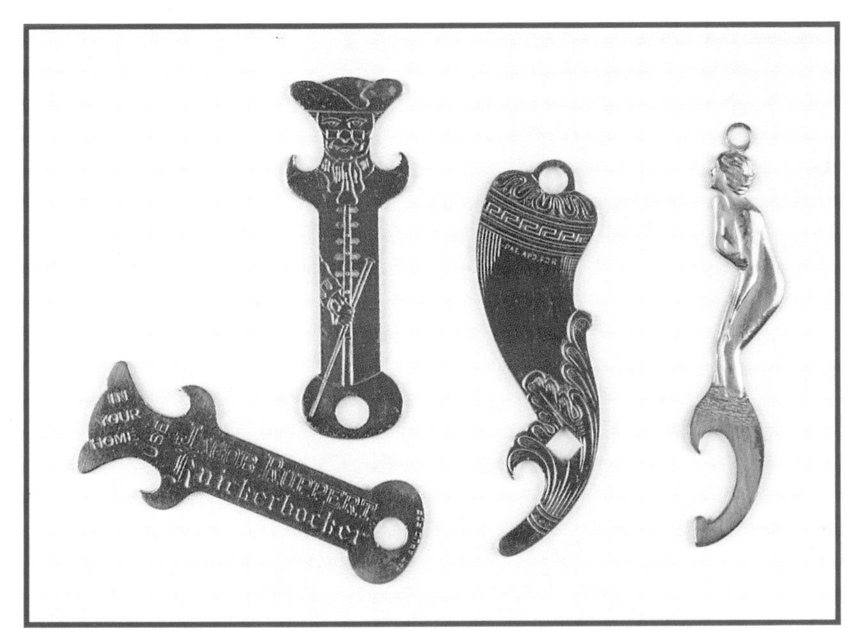

Left pair: A-8 Colonial man. Marked PAT APL'D FOR. 2 3/4". Front and back view. $25-30. *Only found with "In your home use Jacob Ruppert Knickerbocker" advertising.*
Middle: A-34 Powder Horn, Fancy. Prest-O-Lite key. Marked PAT. AP'D. FOR or PATD. 4-28-14. American design patent 45,678 issued to John L. Sommer, April 28, 1914. 3 1/8". $50-100.
Right: A-40 September Morn. Marked COPYRIGHT 1913 BRAUN & CO. SEPT. MORN. High relief brass. 3 1/4". $150-200. *Found with advertising for "Inland Beer" and "Excelsior Beer, Santa Cruz."*

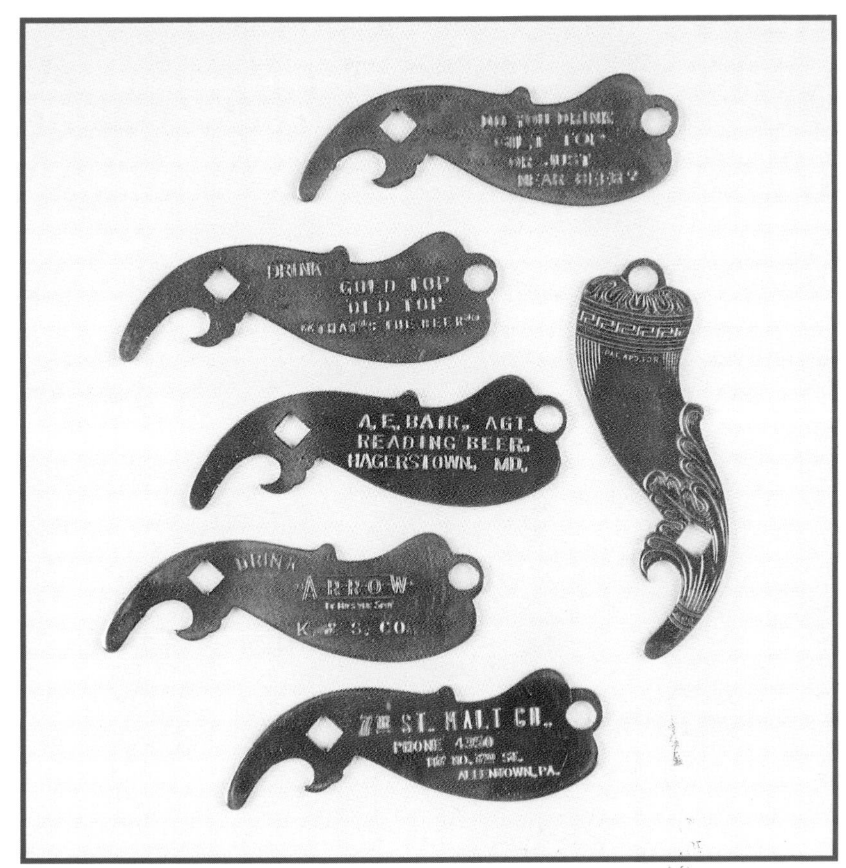

Top to bottom: The five known beer advertising powder horns are shown (with a sixth showing the front view): *"Gilt Top Beer" of Spokane, Washington, "Gold Top Beer" of Columbus, Ohio, "Reading Beer" of Hagerstown, Maryland, "Arrow Beer" of Mishawaka, Indiana, and "7th St. Malt" of Allentown, Pennsylvania.*

September Morn openers depict the image from a painting by the French artist Paul Chabas (1869-1937). He spent three successive summers on the shores of Lake Annecy in Upper Savoy, with a local peasant girl as the model. She was 16 when they started. For the head he is said to have used a sketch of a young American, Julie Phillips, made while she was sitting with her mother in a cafe in Paris. *September Morn* was presented to the New York Metropolitan Museum in 1957 by William Coxe Wright of Philadelphia. Most visitors thought of it as a lovely picture. None reacted as Anthony Comstock did back in May, 1913, when he first saw it in a Manhattan window: "There's too little morning and too much maid. Take it out!" The painting was first exhibited that year in the United States, and Braun & Co. obtained their opener copyright in high relief brass, followed by Harry L. Vaughan's opener type "A-1" in 1914, which was assigned to the Crown Throat and Opener Company of Chicago and came in two styles, clothed and nude.

Left to right: A-6 Lady's leg and shoe. 2 3/4". $15-20. *Found only with advertising: "Compt's West End Brg. Co., Utica-Club, It's in the taste, Pilsener, Wuerzburger-Ginger Ale."*
A-7 Fancy lady's boot. American design patent 42,306 issued to John L. Sommer, March 12, 1912. 3 1/8". $15-30.
A-35 Lady's Boot, Fancy. Prest-O-Lite key. Marked PATD MAR 12-1912. American design patent 42,306 issued to John L. Sommer, March 12, 1912. 3 1/8". $15-35. *The award for most appropriate advertising for the design goes to: "You won't kick if you drink Schooner brew - Bottled by Monroe Rascoe, Reidsville, N.C."*
A-55 Lady's leg. Marked MECO. 3 1/8". $30-40.

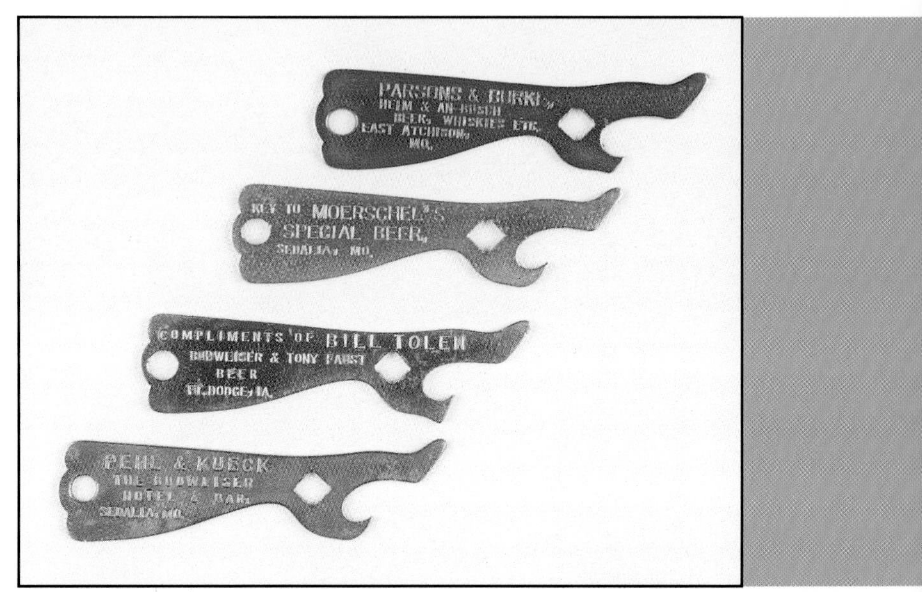

Top to bottom: The four known type A-35 leg openers from Missouri: *"Heim and Anheuser-Busch Beer"* of St. Louis with an Iowa Distributor listed, *"Moerschel's Beer"* of Sedalia, *"Budweiser & Tony Faust Beer"* of St. Louis from another Iowa distributor, and *"The Budweiser Hotel & Bar"* of Sedalia.

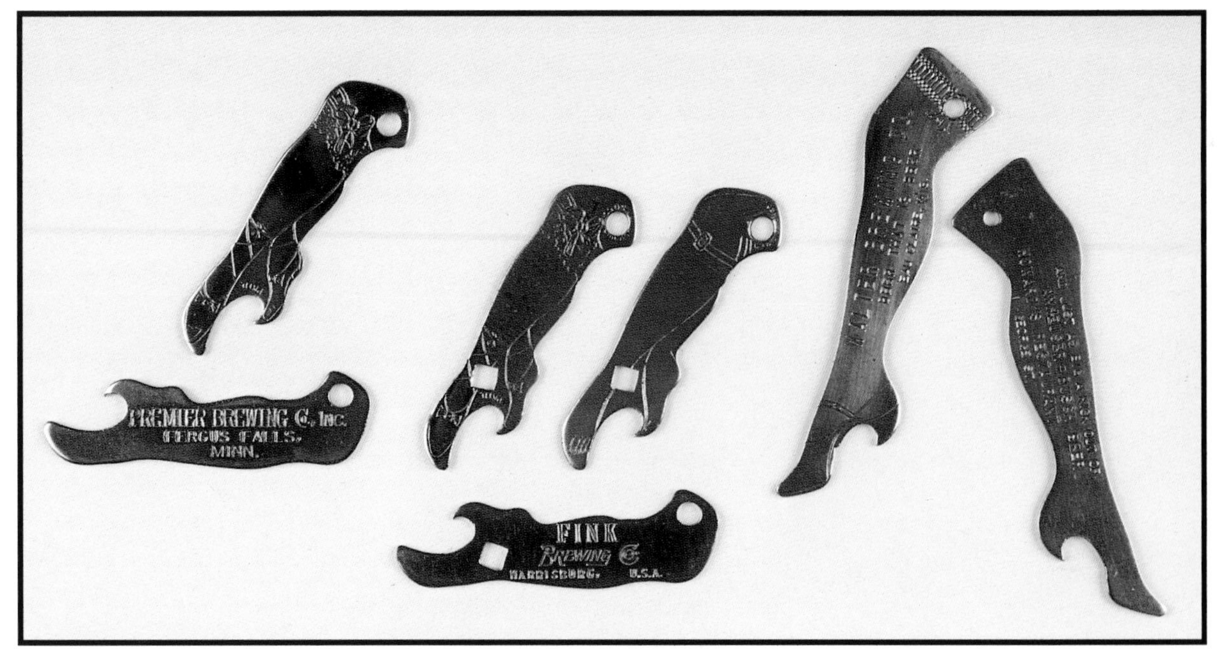

Left pair: A-30 Dancer legs. Called "Dancer" or "Tango." Marked C. T. & O. CO. CHICAGO PATD or PATD. American design patent 44,945 issued to Harry L. Vaughan, November 25, 1913. 2 3/4". $25-50.
Middle three: A-53 Dancer legs with Prest-O-Lite key. Called "Dancer" or "Tango." Marked C. T. & O. CO. CHICAGO PATD or PATD. American design patent 44,945 issued to Harry L. Vaughan, November 25, 1913. 2 3/4". $40-50.
Right pair: A-54 Lady's leg. 4 1/8". $100-125.

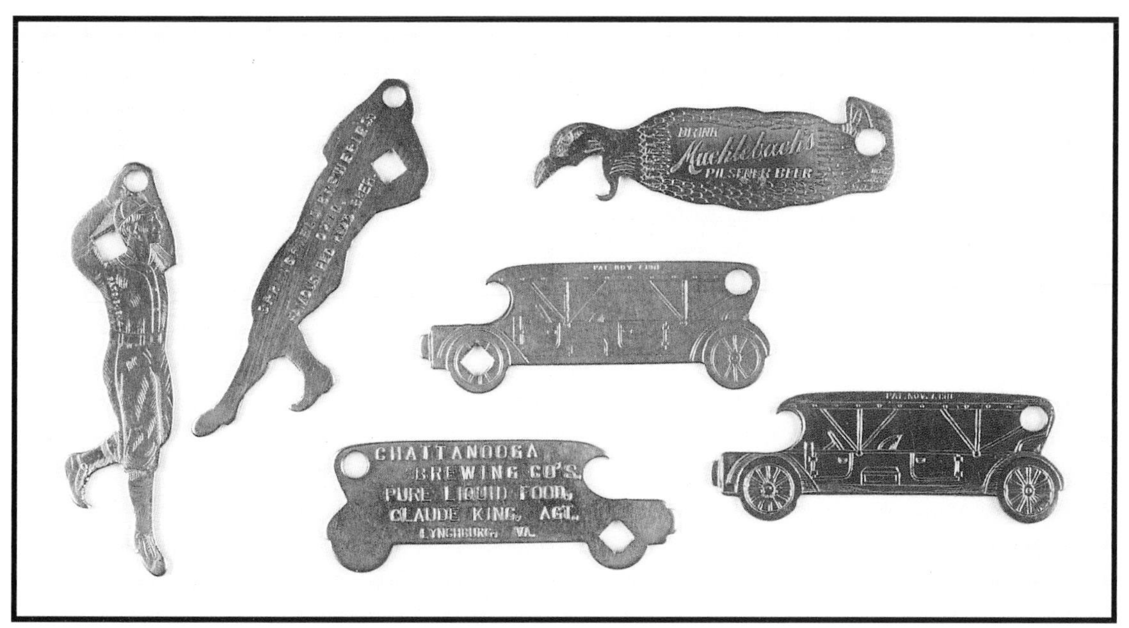

Left pair: A-9 Baseball player in pitching position. Prest-O-Lite key. Marked PATD-8-18-14. One example is marked ADV. NOV. CO. CHICAGO. American design patent 46,298 issued to John L. Sommer, August 18, 1914. 3 1/8". $35-75.
Top right: A-10 Eagle. Marked C. T. & O. CO. Made by Vaughan, Chicago. 2 7/8". $100-125. *Only found with "Drink Muehlebach's Pilsener Beer" advertising.*
Middle pair: A-13 Automobile. Prest-O-Lite key. American design patent 41,895 issued to John L. Sommer, November 7, 1911. 2 7/8". $35-75.
Right: A-52 Automobile. Marked PAT. NOV. 7, 1911. American design patent 41,895 issued to John L. Sommer, November 7, 1911. 2 7/8". $35-50.

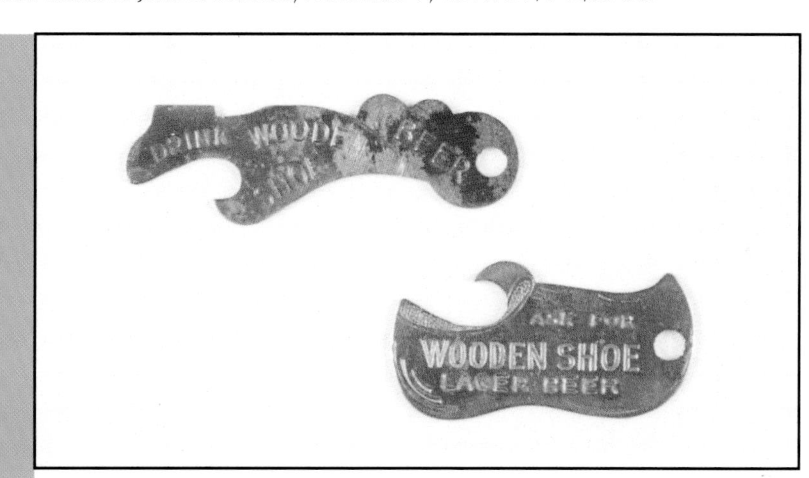

Top: A-11 Shoe. 2 5/8". $75-90. *Only found with "Drink Wooden Shoe Beer" advertising.*
Bottom: A-32 Shoe. 2 1/16". $10-50. *Advertising opener made only for Wooden Shoe Lager Beer, Minster, Ohio.*

Top to bottom:
A-14 Bridge. Copper Plated. 2 3/4". $100-125. *Found only with Rainier Beer and Schmidt Brewing advertising.*
A-19 Alligator. Prest-O-Lite key, cigar box opener, and screwdriver tip. 3 1/8". $125-150.
A-51 Boxing glove. 2". $100-125. *"Blue Ribbon Bouts" (Boxing matches sponsored by Pabst).*

Above right: **Left:** A-12 Sword. 2 7/8". $10-25. *A sword opener from the 1946 Shrine Victory Convention in San Francisco helps date this one.*
Center: A-23 Stainless steel opener made by Louis F. Dow Co., St. Paul, Minnesota. Marked on reverse DOW ST. PAUL STAINLESS. 2 1/2". $10-25. *Advertising to the point appears on one with "This opener will operate best on Schlitz & Pick Ale."*
Pixie pair: A-45 Pixie. 3 1/4". $75-100. *Most appropriate advertising: "If you gotta go, go for City Club, Shorty Billings, Distributor."*

Right: **Top to bottom:** A-15 Eagle Head & Bottle. With and without Prest-O-Lite key. Various marks include C. T. & O. CO., CHICAGO, PAT.APPLD.FOR., C. T.& O. CO., CHICAGO PAT.APPLD.FOR/PATENTED, and C. T. & O. CO. CHGO. PATD 4.30.12. 2 7/8". $15-50.
A-16 Elk or Moose head and bottle. Prest-O-Lite key. Marks include C. T. & O. CO., CHICAGO, PAT APPLD FOR, and C. T. & O. CO., CHICAGO PATENTED. 2 7/8". $25-60. *The Spokane Bottle Supply company, Spokane, Washington, had these imprinted with "Return these keys." An interesting approach to analyzing advertising results.*
A-46 Elk head and bottle. Marked C. T.& O. CO., CHICAGO, PAT APPLD FOR, or C. T. & O. CO., CHICAGO, PATENTED. 2 7/8". $40-60. *Found only with "Drink Rahr's Elk's Head, Oshkosh, Wis." advertising.*
A-17 Lion head and bottle. Prest-O-Lite key. Marks include PATENTED, PAT APPLD, PAT APPL FOR, or PAT. APR.30.12 CROWN T. & O. CO., CHICAGO, ILL. 2 7/8". $15-50. *One example advertises this type as "Vaughan's 'Unique' Pocket Crown Opener."*
A-44 Lion head and bottle. Marked PATENTED. 2 7/8". $20-40.
A-43 Horse head and bottle. Prest-O-Lite key. Marked C. T. & O. CO. PATD. 2 7/8". $50-125.
A-57 Horse head and bottle. Marked C. T. & O. CO. MADE IN U. S. A. 2 7/8". $50-100.

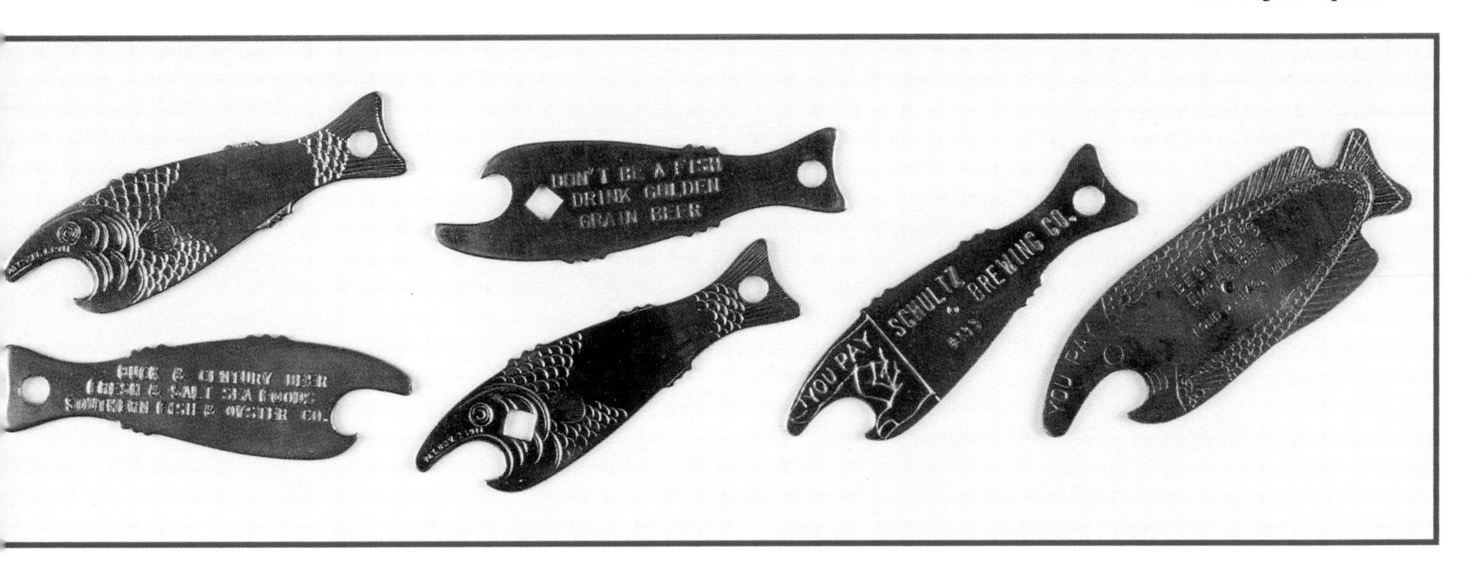

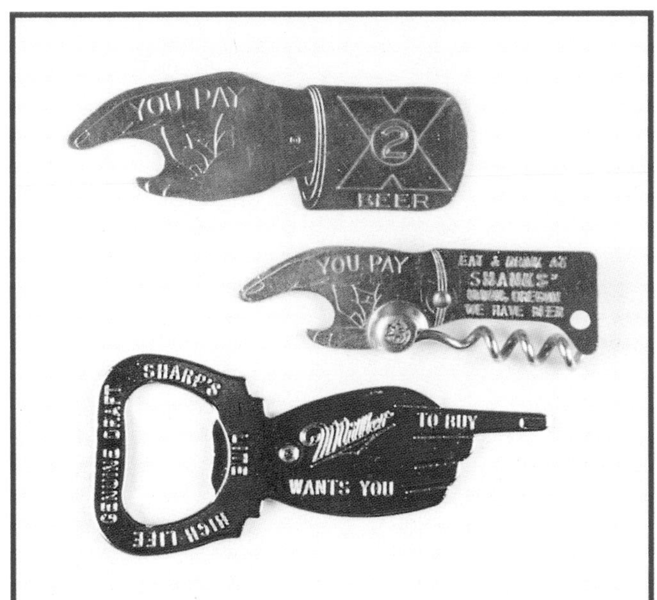

***Above:* Left pair:** A-18 Fish. Marked PAT. NOV. 7, 1911. American design patent 41,894 issued to John L. Sommer, November 7, 1911. 3 1/8". $15-50. **Second pair:** A-42 Fish. Prest-O-Lite key. Marked PAT. NOV. 7, 1911. American design patent 41,894 issued to John L. Sommer, November 7, 1911. 3 1/8". $15-50. *Advertisers "Dubois" and "Golden Grain" used the slogan "Don't be a fish, drink (their brand name)."* **Second from right:** A-20 Fish Spinner. There is a knob punched in the center. The opener spins on the knob and whomever it points to, pays for the drinks. 3 1/8". $10-35. *This opener has been found with this copy: "Advertising specialties thru Chicago office J. L. Sommer Mfg. Co., 604 Hearst Bldg., C. J. Kaufmann, 2415 No. 18th St., Milwaukee, Wis."* **Right:** A-33 Fish, large. Spinner made by L. F. Dow Co., St. Paul, Minnesota. 3 1/4". $50-125.

***Center:* Top:** A-21 Hand Spinner. Usually marked SPIN TO SEE WHO PAYS, MADE BY BROWN & BIGELOW CO., ST. PAUL, MINN. 3 1/4". $2-25. *One advertiser includes the copy "Spin for a bottle of premium quality Falstaff." The Jack Daniels Brewery in Lynchburg, Tennessee, currently offers a hand spinner for $3.00, advertising their "1866 Classic Oak-Aged Beers."*
Middle: A-36 Hand spinner with corkscrew. 3". $75-100. *None have been reported with beer advertising. One advertiser is: "Triner's wines, cordials, liquors."*
Bottom: A-56 Hand spinner. 3 5/8". $10-15. *Advertising only for Miller.*

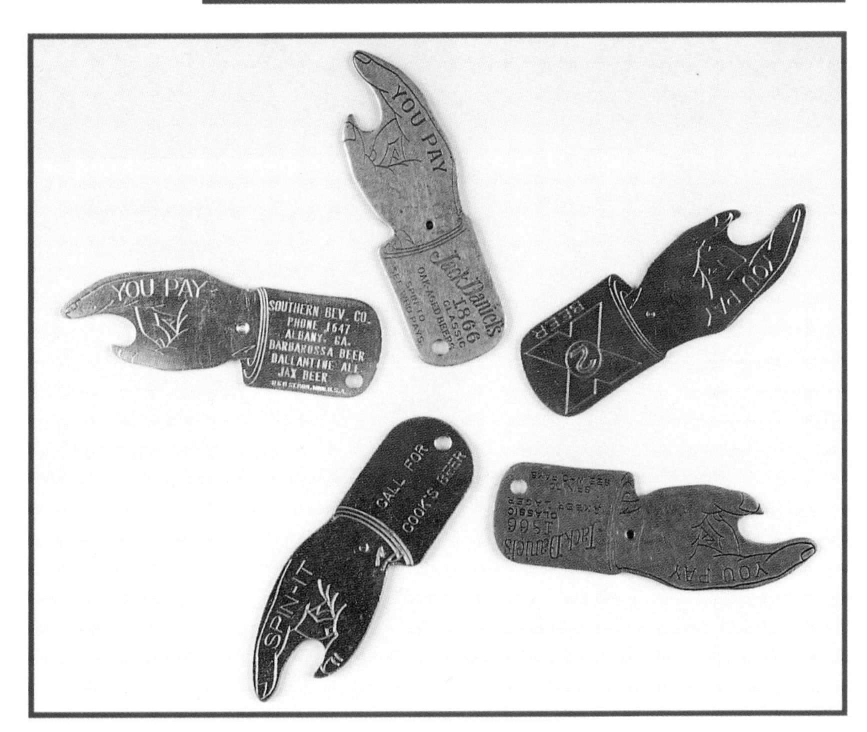

Bottom: Five type A-21 spinning hands: aluminum from *"Southern Beverage Co.;"* "Spin-It" advertising *"Cook's Beer;"* "You-Pay" advertising *"2 X Beer"* from Texas; and recently made examples by the Jack Daniels brewery.

Left: A-22 Bottle. 2 7/8". $125-150.
Middle: A-24 Bottle, enameled. 2 1/2". $35-75. *The manufacturer used this to advertise their own products with "Made by Etching Co. of America, Chicago, Ill., Etched metal signs and specialties, bottle opener."*
Right: A-27 Bottle. Extra thick with Prest-O-Lite key and screwdriver tip. 3 1/4". $60-75.

Left pair: A-25 Bottle, enameled. 2 7/8" to 3". $25-100.
Middle pair: A-26 Bottle, enameled. 3 1/2". $25-100.
Right pair: A-38 Bottle, enameled. 2 3/4". $25-100.

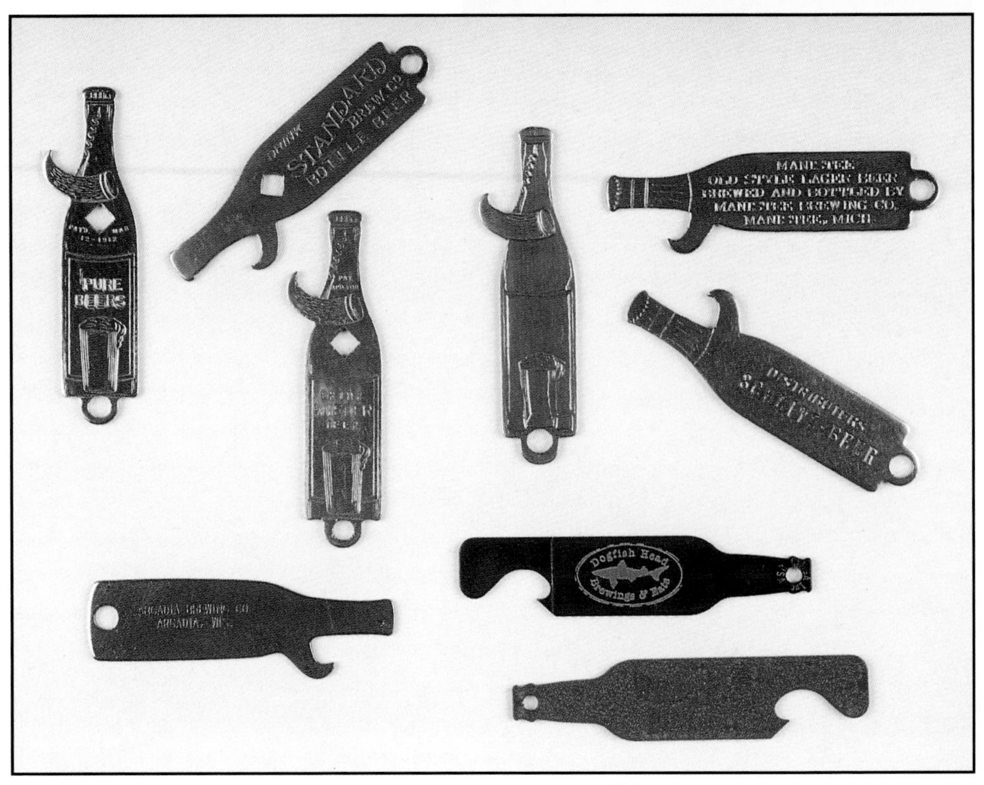

Top left three: A-28 Bottle. Pictures a stag handle corkscrew. Some show a glass of beer. Prest-O-Lite key. Marked PATD. MAR. 12-1912 or PAT APD FOR. American design patent 42,305 issued to John L. Sommer, March 12, 1912. 3 1/4". $10-35.
Top right three: A-29 Bottle. Some picture a stag handle corkscrew. Some show a glass of beer. Marked PATD. MAR. 12-1912 or PAT APD FOR. American design patent 42,305 issued to John L. Sommer, March 12, 1912. 3 1/4". $5-35.
Bottom left: A-50 Bottle. 3". $35-40.
Bottom right pair: A-58 Bottle. Manufactured by Cymba, Inc., Connecticut. 3 1/2". $3-5.

A-63 Bottle made of Titanium. Made expressly for Anchor Steam of San Francisco. 2 3/4". $8-10.

Left: A-37 Hand. Prest-O-Lite key. Along with A-59 and A-60 one of the Rhode Island "Big-Three Figurals." 3 1/2". $125-150. *Found only with "Hand Brewery" advertising.*
Middle: A-59 Hand holding bottle. Marked U. F. & S. CO. 3 3/8". $150-200. *Found only with "Compliments of Providence Brewing Co. Providence, R. I. U.S.A./ Export Bohemian Beer" advertising.*
Right: A-60 Indian headdress. Prest-O-Lite key. 3". $150-200. *Found only with "Famous Narragansett Lager, Banquet Ale" advertising.*

Left: A-31 Seal. 3 3/8". $20-25. *Has only been found with advertising "Get the National habit, cash in on National Oil Seals." May not have been produced with other advertising.*
Middle pair: A-39 Turtle with three screwdrivers. Marked PATD 161,321 B&B U.S.A. Made by Brown & Bigelow, St. Paul, Minnesota. American design patent 161,321 issued to Le Emmette V. De Fee, December 26, 1950. 3". $40-50.
Right: A-48 Turtle with three screwdrivers. Plastic covered steel body. Marked LEWTAN U.S.A. PAT. PEND. 2 3/4". $20-25.

Left: A-41 Dog. 3 3/4". $75-100. *Found only with advertising for "Medford Lager Beer, Medford, Wis."*
Right: A-61 Donkey. 3". $100-125.
Bottom: A-62 Race car. Marked COPYRIGHT COOL PRODUCTS. 6 1/4". $5-8.

Top: A-47 Eagle on shield. Key Holder. 3". $125-150. *Found only with advertising: "Fred Krug Brewing Co., Omaha, U. S. A., Key to Luxus, the beer you like, Grand Aerie Convention, Omaha, U. S. A., 1909."*
Bottom: A-49 Eagle. Prest-O-Lite key and key holder. 3 1/8". $125-150. *Found only with advertising: "Gilt Top Beer Spokane, Wash."*

Key Shape Openers

Although very few of the openers in this chapter look like a key, they all have a hole or other provision for adding to a key chain. And they are the "key" to opening a bottle. Advertising frequently includes the words "Key to" before the beer brand name. Some have been found in leather cases in which the opener and keys swivel out on a pin for use. An interesting notice was included by one advertising specialty company on this type: "Finder rewarded if keys are returned or drop in any mail box. L. F. Grammes & Sons, Allentown, Pa." The owner's registration number was engraved on the opener. Most of the types were produced and used prior to the 1919-1933 Prohibition period.

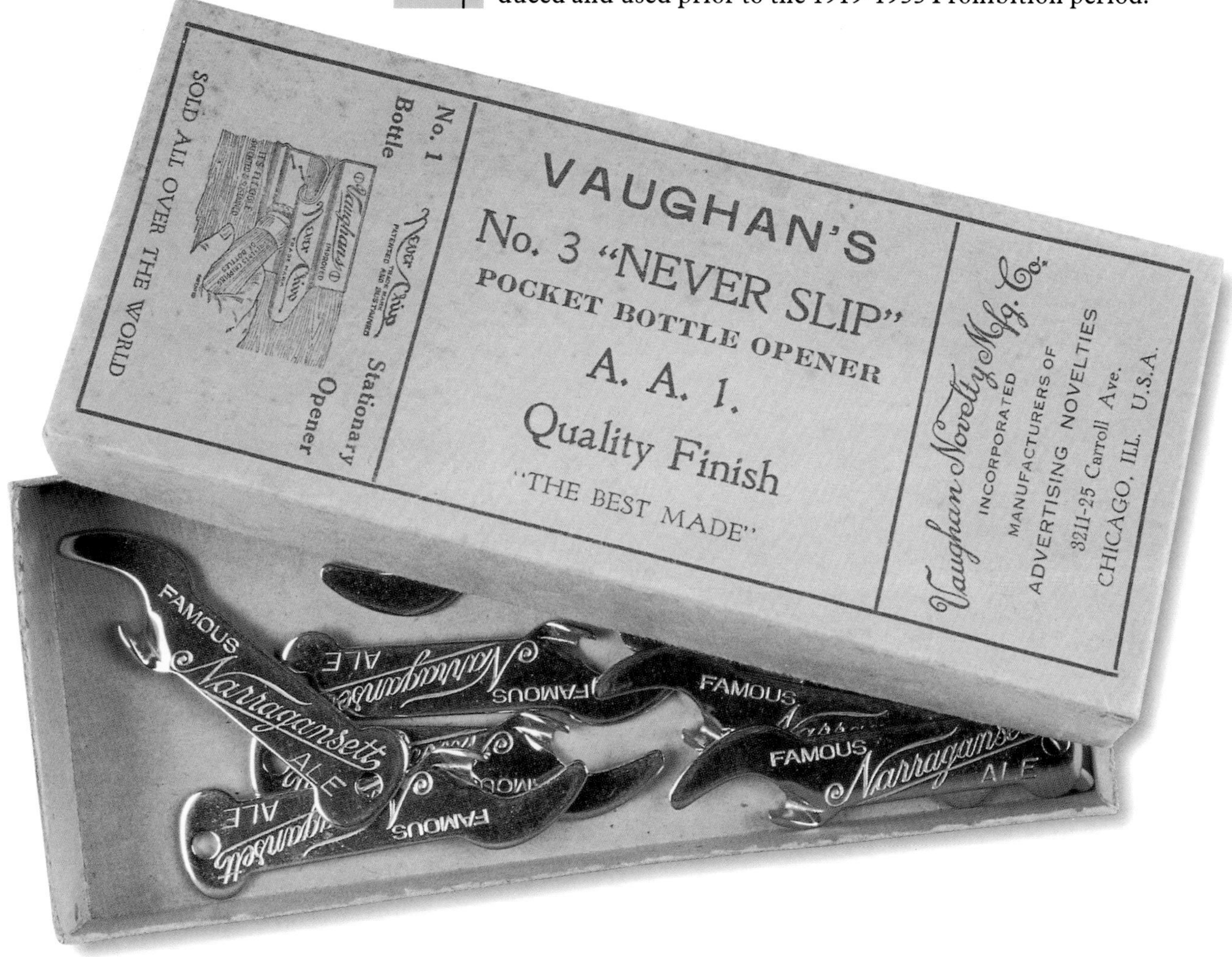

***Top:* Top to bottom:**B-1 Brass or steel enameled cap lifter. 2 1/2". $20-75. *Found with advertising for the manufacturer: "Electro Chemical Engraving Co., Suite 511, 90 West St., New York City, Highest Quality."*
B-2 Brass or steel enameled cap lifter. Prest-O-Lite key. 3 1/8". $20-75. *J. K. Aldrich, 1098 Brook Avenue, New York used this type to advertise "Bottle Openers" and "Advertising Metal Material."*
B-3 Brass or steel enameled cap lifter. Prest-O-Lite key. Cigar box opener. Nail puller. 3 1/8". $35-75.
B-4 Brass or steel enameled cap lifter. 3 1/8". $10-35.
B-33 Cap lifter, enameled brass or steel. Key chain hole larger than B-4 and B-60. 3 1/8". $20-30.
B-60 Cap lifter. Enameled brass or steel. 3 1/8". $30-75. *One advertising Stark-Tuscarawas Breweries Company, Canton, Ohio, celebrated the "Al Koran Shriners, Sept. 10, 1914."*

Below: Several varieties of type B-1. These openers are very appealing because of company logos usually added.
Left: *"Rising Sun"* of Elizabeth, New Jersey; *"Forest City"* of Cleveland, Ohio; *"Ben Brew"* from Franklin Brewing Co. of Columbus, Ohio; *"Burkhardt"* of Akron Ohio.
Middle: *"Kauffman"* of Cincinnati, Ohio; *"Hoster-Columbus"* of Columbus, Ohio; *"C & J Michel"* of La Crosse, Wisconsin.
Right: *"Ideal Beer"* of Bridgeport, Connecticut; *"Wagner Beer"* of Sydney, Ohio; *"Susquehanna Beer"* of Nanticoke, Pennsylvania; and, finally, a crown cap go-with by *"Hutchinson Crowns"* of Chicago, Illinois.

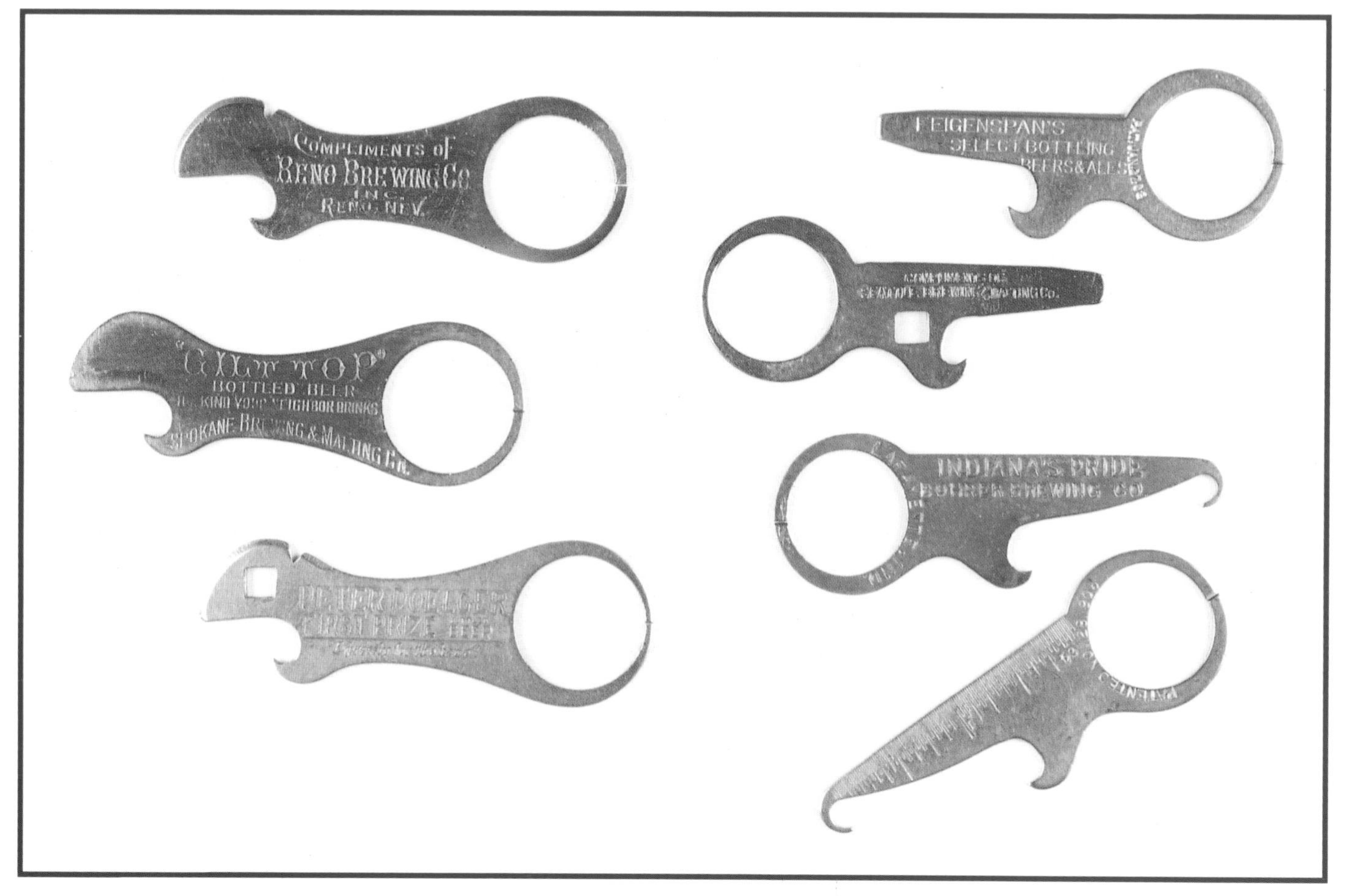

Top left: B-5 Combination cap lifter, cigar box opener, nail puller, and key holder. Marked PAT APL'D FOR. 3 1/8". $25-60. *A clue to the source comes from one with advertising: "Compliments of The Western Brewer, H. S. Rich & Co., Publishers, Chicago - New York, Manufactured by National Selling Co., Allentown, Pa."*

Middle left: B-26 Combination cap lifter, key holder, and cigar box opener. 3 1/8". $40-60. *One with advertising for Spokane Brewing & Malting is marked MANUFACTURED BY NATIONAL SELLING CO., ALLENTOWN, PA.*

Bottom left: B-42 Combination cap lifter with cigar box opener, nail puller, key holder, and Prest-O-Lite key. Marked PAT APL'D FOR. 3 1/8". $25-50.

Top right: B-6 Combination cap lifter, screwdriver, and key holder. Some marked PAT. JAN. 27, 03. Made by Whitehead & Hoag Co., Newark, New Jersey. 2 7/8". $20-50. *Also found with an advertising novelty distributor's message "G. P. Coates Co., Mfrs. of Original Ad. Novelties, Norwich, Conn."*

Middle right: B-46 Combination cap lifter with Prest-O-Lite key, screwdriver, and key holder. Made by Whitehead & Hoag Co., Newark, New Jersey. Some marked PAT. JAN. 27, 03. 2 7/8". $40-50.

Bottom right pair: B-7 Combination cap lifter, key holder, button hook, and ruler. Marked PATENTED NOV. 28, 1905, and sometimes including HANDY POCKET COMPANION. American patent 805,486 issued to Julius T. Rosenheimer, November 28, 1905. 3 1/8". $30-60. *The Handy Pocket Companion had eighteen uses: Bottle cap opener, key ring, ruler, lifting and pulling carpet tacks, lifting pots or kettles from fire or stove, glove buttoner, cigar box opener, pulling rusty pens from holder, winding or tightening window shade springs, shoe buttoner, pulling wire from bottles, watch case opener, nail file and cleaner, basting thread puller, paint can opener, and wide mouth bottle opener. What more could you ask for?*

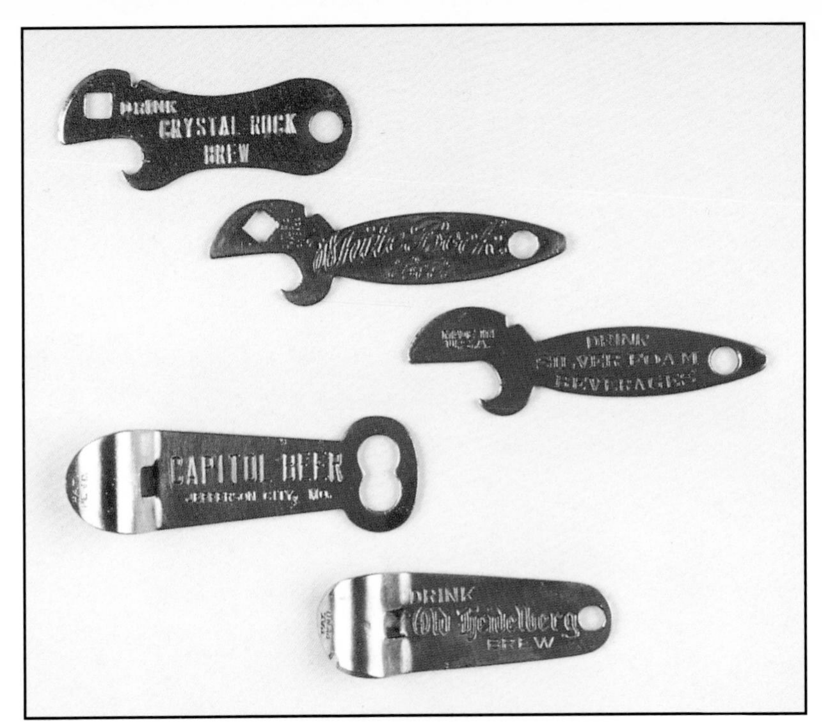

Top to bottom: B-8 Combination cap lifter, Prest-O-Lite key, cigar box opener, and nail puller. Made by L. F. Grammes & Sons of Allentown, Pennsylvania. 2 3/8". $25-60. B-9 Combination cap lifter, Prest-O-Lite key, cigar box opener, and nail puller. Cigar shape. Marked C.T.&O. CO. PATD. CHICAGO or PAT.APL FOR. 2 7/8". $25-75.B-65 Cap lifter with cigar box opener and nail puller. Cigar shape. 2 7/8". $30-40.B-30 Cap lifter with double punched key chain hole. Stamping with formed end. 2 3/4". $20-40.B-49 Cap lifter with cigar box opener and nail puller. Stamping with formed end. 2 1/4". $40-50.

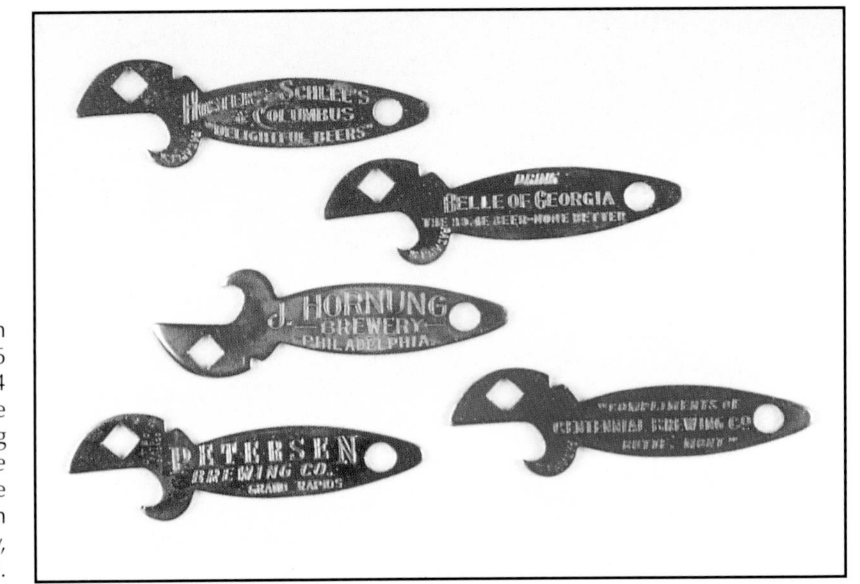

Five examples of Type B-9, Crown Throat & Opener Company's "No. 6 Cigar" opener shown in a 1914 advertisement. Ads are very descriptive in the examples with the Hornung being the back side of the *"White Bock"* opener shown as type B-9. The *"Belle Of Georgia"* has advertising from the Augusta Brewing Company, Augusta, Georgia, on the back side.

Top one: B-10 Cap lifter with Prest-O-Lite key. Marked PAT APLD FOR. 2 1/2". $75-100. **The rest:** B-29 Cap lifter with Prest-O-Lite key. 3". $30-75. *This was Williamson Company's catalog number 108 advertised as "Prest-O-Lite tank key & crown opener."*

On July 17, 1911, six days of excitement began on the docks in Seattle. The event had been dubbed "Golden Potlatch" by Seattle school teacher Pearl Dortt. She had read of the custom among Indians of bringing and hoarding their treasure for many years and when they had gathered a great supply, they invited friends and members of their tribe to celebrate giving their treasures away. The custom was known as "Potlatch." The "Golden" was added as the six day festival was in commemoration of the arrival of the first gold ship from Alaska. During this festival large firms occupied booths where they gave away small and useful articles publicizing or advertising their products. Seattle Brewing and Malting Company, which then produced Rainier Beer, had such a booth. Every hour on the hour they gave away a very limited number of beer bottle openers. These had a "Potlatch" bug set in enamel and embossed with gold on the handle. People stood in lines at the Seattle Brewing booth, some of them for several hours, to get one of these openers. The Golden Potlatch festivals were subsequently held in 1912, 1913, 1914, 1934, 1938, 1939, 1940, and 1941.

The "standard" Potlatch symbol with a spire and question mark was in wide use on various souvenirs. An explanation from an old Seattle antique dealer for the spire and question mark: "In 1911, a proposal was made to build in Seattle the tallest building 'west of the Mississippi'. There was considerable doubt that it would ever be built. This was reflected on the Rainier Potlatch symbol. However, it was built in 1913 and today is still the favorite building of the majority of Seattleites despite being dwarfed by numerous other ones. The Smith Tower has a top like a Potlatch symbol. The opener with a spire and question mark was probably given out in 1911 and 1912. After completion of the L.C. Smith building, in 1913 and 1914, the 'Potlatch' symbol opener which faces in the opposite direction was handed out. It seems that both openers are equally scarce and desired by collectors.

Left: **Top three:** B-11 Cap lifter, enameled on gold. "Potlatch" or totem pole figure attached. 3". $75-125. *Believed to have been distributed at an Alaskan-Yukon Exposition by Seattle Brewing and Malting Company.* **Bottom four:** B-27 Cap lifter, enameled brass or steel. May be marked THE GREENDUCK CO., CHICAGO. 2 1/2". $35-75. *J. K. Aldrich, 1098 Brook Avenue, New York used this type to advertise "Bottle Openers."*

Below: **Left side, top to bottom:** B-12 Cap lifter with folding button hook. Marked VAUGHAN PATD CHICAGO. 2 7/8". $50-75.
B-13 Cap lifter with folding corkscrew known as the "Nifty." Marked MADE & PAT'D IN U.S.A. VAUGHAN, CHICAGO. Double steel body. American patent 1,207,100 issued to Harry L. Vaughan, December 5, 1916. 2 7/8". $10-50.
B-35 Cap lifter with folding corkscrew. Single steel body. Some marked PAT. NO. 1680291. Made by John L. Sommer Manufacturing Co. American patent 1,680,291 issued to Thomas Harding, August 14, 1928. 2 7/8". $15-35. *Classic advertising on one from Schneider Brewing Company, Trinidad, Colorado, reads "Century Beer, the brew you'll enjoy a century in every bottle."* **Right side, top to bottom:**
B-36 Cap lifter with folding corkscrew. Single steel body. Cap lifter and corkscrew are on opposite sides. 2 7/8". $20-30.
B-38 Cap lifter with folding corkscrew. Prest-O-Lite key where key chain hole appears on others. 2 7/8". $20-30.
B-48 Cap lifter with folding screwdriver. Double steel body. 2 7/8". $50-65.

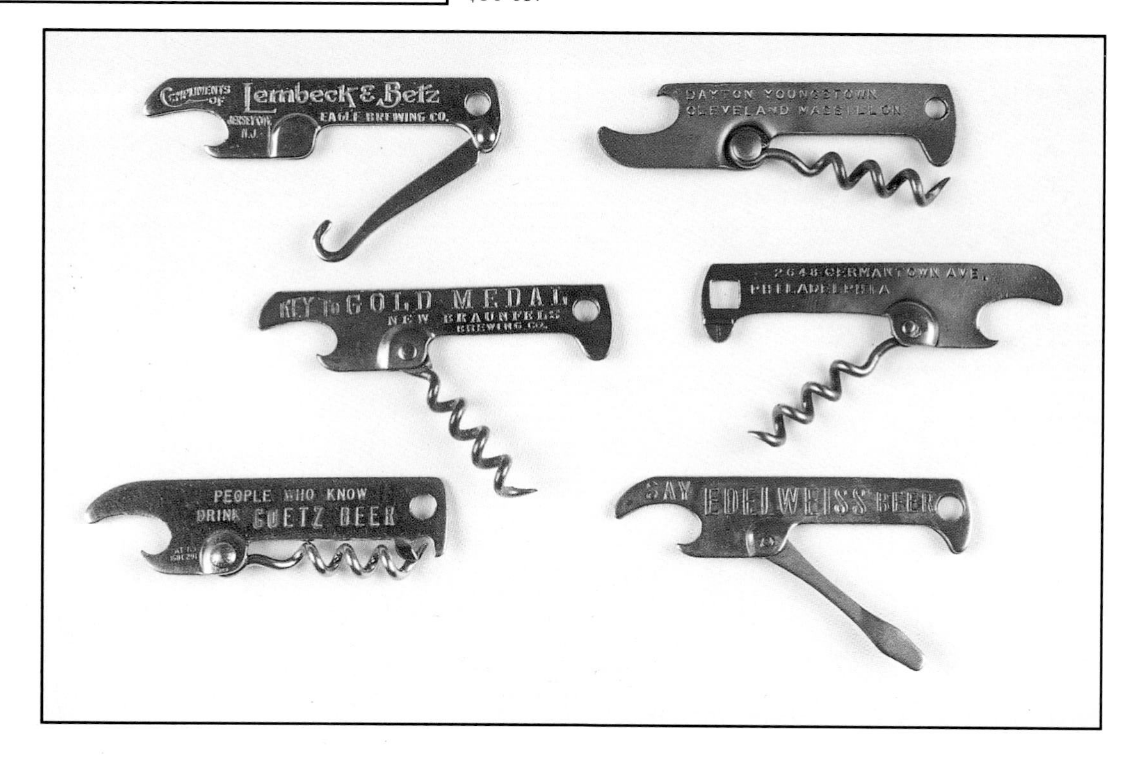

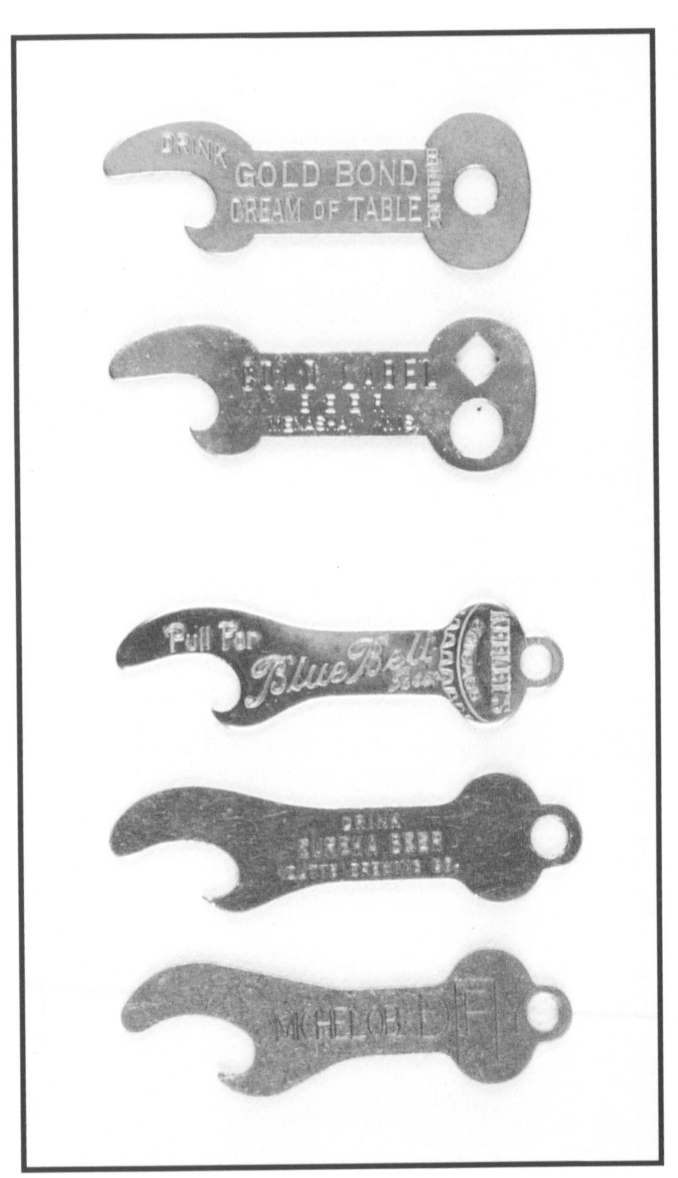

Top to bottom: B-15 Cap lifter. 2 1/2". $10-30.
B-16 Cap lifter with Prest-O-Lite key. 2 1/2". $20-40.
B-14 Vaughan's "Special Pocket Bottle Opener." 2 5/8". $3-25. *A B-14 from an advertising specialty company says "Improved Crown & Seal Co., Made in Chicago."*
B-41 Cap lifter. 2 3/4". $30-40. *Found with advertising from the manufacturer: "Louis F. Dow Co., St. Paul, Minn., No. 100 Bottle Opener."*
B-69 Cap lifter. 2 5/8". $10-15.

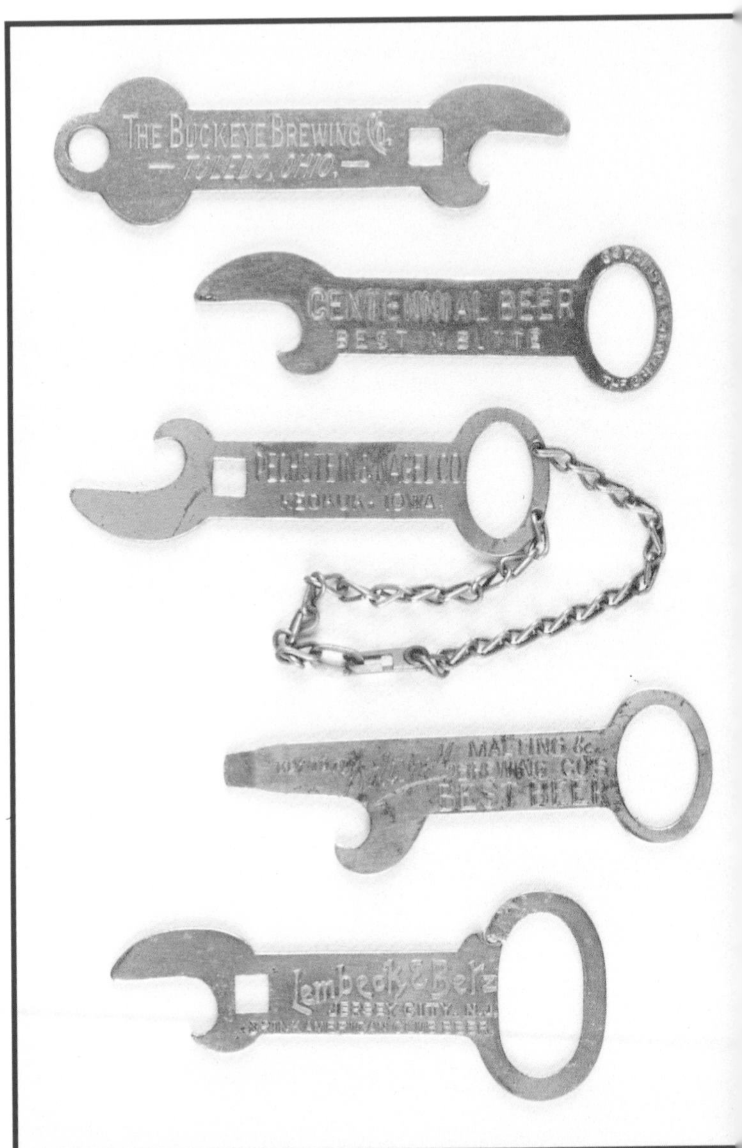

Top to bottom: B-25 Cap lifter with Prest-O-Lite key. 3 1/4". $25-40.
B-17 Cap lifter with oval key chain hole. Marked THE GREENDUCK CO., CHICAGO. 3". $25-50.
B-34 Cap lifter with oval key chain hole. Prest-O-Lite key. Marked THE GREENDUCK CO., CHICAGO. 3". $60-75.
B-43 Cap lifter with screwdriver tip and oval key chain hole. 3". $100-125.
B-44 Cap lifter. Prest-O-Lite key and oversized key holder. 3 1/4". $100-125.

Note: American design patent 34,096 was issued to Augustus W. Stephens, Feb. 19, 1901. This patent date appears on many of the type B-18, B-19, B-21, B-52, and B-63 openers. Some are marked PICNIC - REG. U.S.PAT.OFF. TRADE MARK and PAT. FEB. 19, 1901. In addition, various manufacturer or distributor names appear on them. These include:

Adv. Novelty Co., Chicago
Bachrach & Co., San Francisco
E. M. Blumenthal & Co., Chicago
Thomas Burdette, Montreal, P. Q.
Hugo Cahn & Co., New York
Chicago Spec. Box Co.
Colson Co.
Oscar Heyman & Co., 43 Park Place, N. Y. C.
A. Magnus Sons Co., Chicago, Ill.
Quimby Mfg. Co., Minneapolis, Minn.
M. C. Rosenfeld Co., Boston
A. W. Stephens Mfg. Co., Cambridge, Mass.
A. W. Stephens Mfg. Co., Waltham, Mass.
Witteman Bros., New York
and currently made by Cymba, Inc.

Manufacturer and distributor advertisements on these types include:

- This opener stamped to order by Chicago Specialty Box Co. makes a fine advertisement.
- This is a Picnic Crown Bottle Opener (A. W. Stephens Co.).
- Send for prices, No. 106 (Williamson, Newark, New Jersey).
- Williamson hard rubber finish No. 107, New Ready Crown Opener.
- The Presto Bottle Opener and Gas Tank Key.
- Compliments of A. W. Stephens Mfg. Co., Originators of this bottle opener.

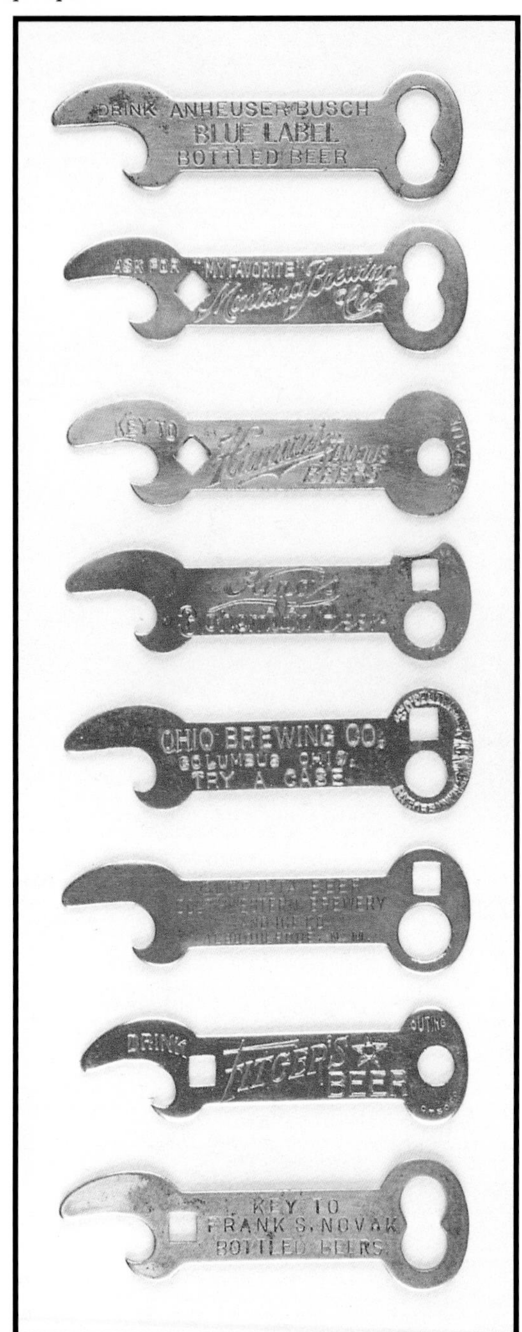

Top to bottom: B-18 Cap lifter with double punched key chain hole. Stephens design patent of 1901. 3". $3-35. *Olympia Brewing Company of Olympia, Washington, had 10,000 of these made with advertising "Return these keys to Olympia Brewing Co., Olympia, Washington, and receive reward." They were numbered 1 through 10,000. A commemorative example reads "Membership key to Schwarzenbach Brewing Co., Hornell, N. Y., C. N. Y. V. Firemen's Ass'n, Hornell, N. Y., July 20-21-22, 1909."* B-19 Cap lifter with double punched key chain hole. Diamond Prest-O-Lite key. Stephens design patent of 1901. 3". $3-35. B-47 Cap lifter with Prest-O-Lite key (diamond). 2 7/8". $25-35. B-20 Cap lifter with Prest-O-Lite key. Quarter punch missing above gas key. 3". $25-30. B-21 Cap lifter with Prest-O-Lite key. Stephens design patent of 1901. 3". $3-35. *One with an interesting bit of advertising reads "Briggs Ale 'The kind that won't ferment in your stomach'." Both Olympia Brewing Company of Washington and Park Brew Company of Providence, Rhode Island, had this type produced and numbered 1 through 10,000. The slogan of Schlitz, Milwaukee, "The beer that made Milwaukee famous" was well used and appeared on many opener types. Schwarzenbach Brewing Company of Hornell, New York, one-upped Schlitz with the advertising slogan on a B-21 type opener—"The beer that made itself famous!"* B-63 Cap lifter with Prest-O-Lite key. Much larger key chain hole than B-21. 3". $15-35. B-22 Cap lifter with Prest-O-Lite key. Vaughan's pre-prohibition "Outing" key type pocket bottle opener. 2 7/8". $3-35. *Another place made famous: "The Kuntz Brewery Limited, Call for Kuntz, the beer that made Waterloo famous" (Canada).* B-52 Cap lifter with Prest-O-Lite key. Double punched key chain hole. 3". $30-40.

Top to bottom: B-24 Vaughan's "Never Slip" bottle opener. American patent 2,018,083 issued to James A. Murdock, October 22, 1935. 3 1/8". $2-30. *Vaughan proclaimed on one of these "World's largest manufacturers of bottle openers. Compliments of Vaughan Novelty Mfg. Co., Chicago, Ill."* B-23 Cap lifter. Vaughan's post-prohibition "Outing" key type pocket bottle opener. 2 7/8". A second example with a spinner button is also shown. $3-30. *One thousand were numbered for this advertisement: "Return these keys to Salem Brewery Association, Salem, Oregon, and receive reward."* B-64 Cap lifter. Much larger key chain hole than B-23. 2 7/8". $15-25. B-54 Cap lifter. 3". $30-40. B-68 Cap lifter with screwdriver tip. 3 1/8". $50-60.

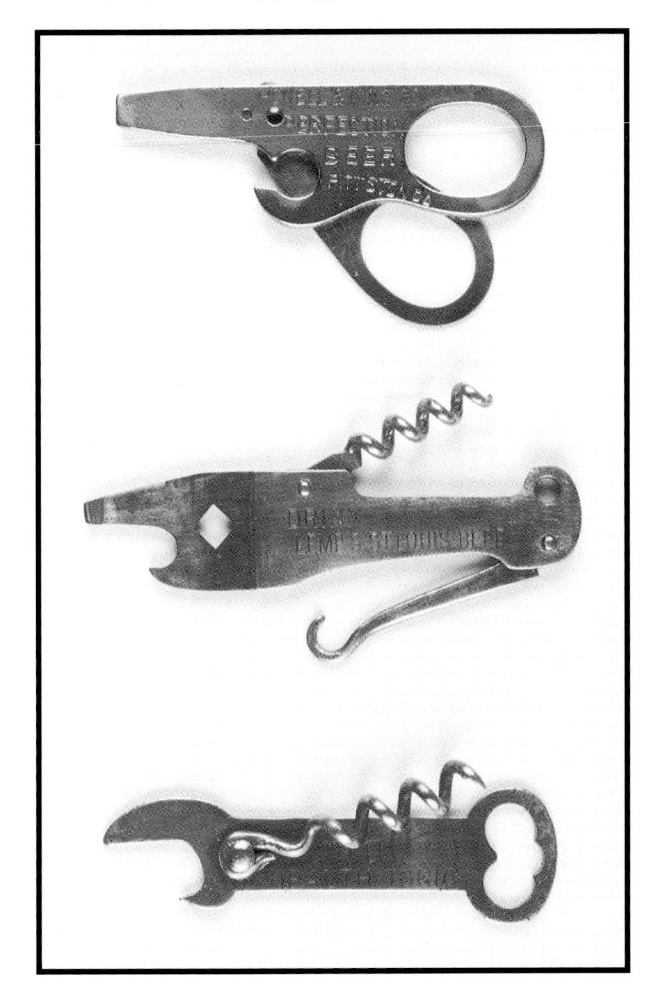

Top to bottom: B-28 Cap lifter with screw driver and cigar cutter. Made by E & D Mfg. Co., of Michigan. American patent 1,124,288 issued to Elvah V. Bulman, January 12, 1915. 2 7/8". $125-150. B-45 Cap lifter with Prest-O-Lite key, folding corkscrew, folding button hook, and screwdriver tip. 3 3/8". $150-200. *An advertising specialty company used this type for its message: "Metal States Novelty Co., Salt Lake City, Pat. Pending."* B-66 Cap lifter with folding corkscrew. 3". $50-60.

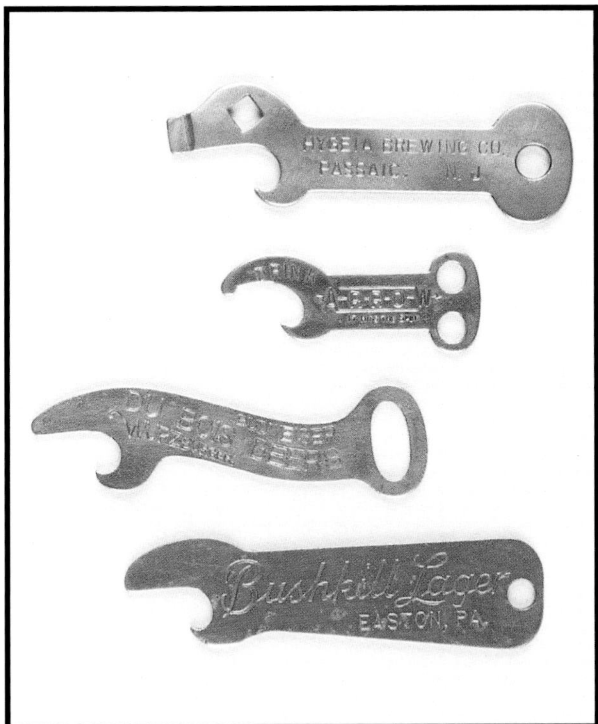

Top to bottom: B-31 Cap lifter with Prest-O-Lite key. Screwdriver tip. Extra thick steel. 3 1/4". $30-50. *This was Williamson Company's catalog number 110 advertised as "Prest-O-Lite tank key, crown opener, screw driver."* B-32 Cap lifter with two key chain holes. 2". $75-100. B-37 Cap lifter with oval key chain hole. Wave form. 3". $75-100. B-39 Cap lifter. 3 1/4". $60-75. Made by L.F. Grammes & Sons of Allentown, Pennsylvania. *Found only with "Bushkill Lager, Easton, Pa." advertising.*

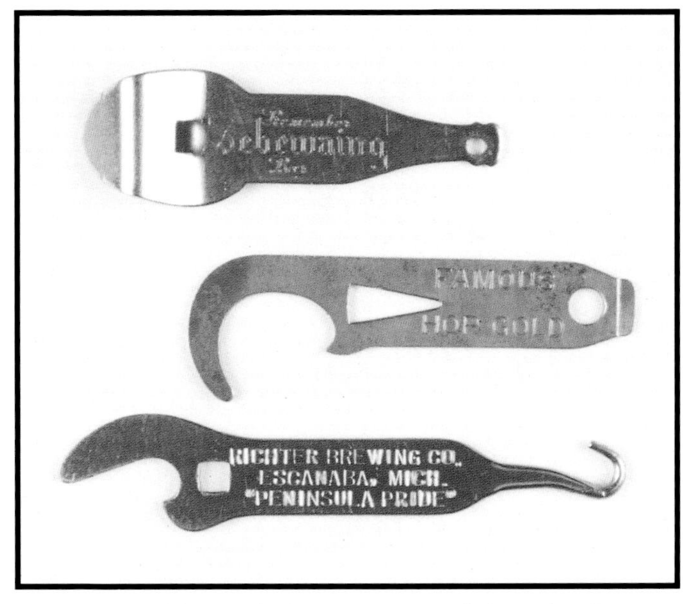

Top: B-40 Cap lifter. Bottle shape stamping with formed end. Marked EMRO ST.L. MO. NOT FOR RESALE. 2 7/8". $75-100. *A simple reminder appears on the only known beer advertising opener of this type: "Remember Sebewaing Beer."* **Middle:** B-50 Cap lifter with screwdriver tip and nail puller. 3 1/8". $50-60. **Bottom:** B-53 Cap lifter with Prest-O-Lite key and button hook. 3 3/4". $60-75.

Top: B-70 Folding cap lifter with Prest-O-Lite key and screwdriver tip. Marked MFD BY INDIVIDUAL KEY RING CO. HARTFORD CT. PAT APLD FOR. American patent 1,040,564 for "Composite Tool" issued to A. W. Merrill, October 6, 1912. 2 3/4". $100-125.
Middle: B-71 Folding cap lifter with three key chain holes. 2 3/8" closed. 3" open. $100-150.
Bottom: B-72 Cap lifter with rotating cigar cutter, key holder, and screwdriver tip. Marked PAT. APLD FOR. 3". $100-125.

Top from left to right: B-55 Cap lifter with applied advertising button. 2 1/4". $1-3.
B-57 Cap lifter with plastic insert. Marked MADE IN HONG KONG. 2 3/8". $1-3.
B-58 Cap lifter with plastic insert. 2". $1-3.
B-62 Cap lifter with plastic insert. 2 3/8". $1-3.
B-73 Cap lifter with plastic insert. 2 3/8". $3-5.
Bottom: B-74 Cap lifter in shape of large key. Manufactured by Cymba, Inc. 3 3/8". $8-10.

Flat Metal Cap Lifters

This type of opener is usually a metal stamping with a large hole that has a tab to engage the underside of a crown cap. The handle is lifted up to remove the cap. In some instances a second tab is found at the top of the opening, enabling the user to remove the cap by engaging the lip opposite the handle. The cap is removed by pushing the handle down. The openers are found with vertical and horizontal advertising. The horizontal advertising is most common on right-handers and less frequently seen on left-handers. When placed in the right-hand to open a bottle, the advertising on a left-hand model will be upside down (and vice-versa).

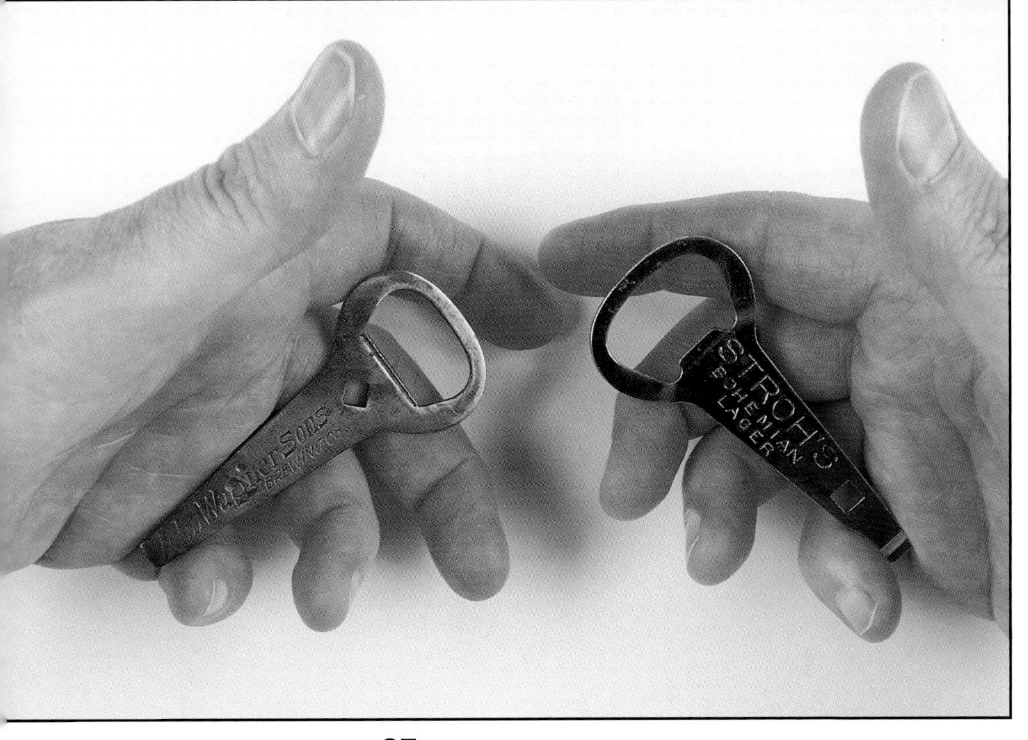

Top: C-1 Cap lifter with raised letters and enameled background. Marked PATENT APP FOR or PATENTED FEB.16,1916. American design patent 48,550 issued to Nelson Jacobus, February 16, 1916. 2 1/2". $35-100.
Bottom: C-2 Cap lifter with raised letters and enameled background. Marked PATENTED FEB.16, 1916. 3 3/8". $35-100.

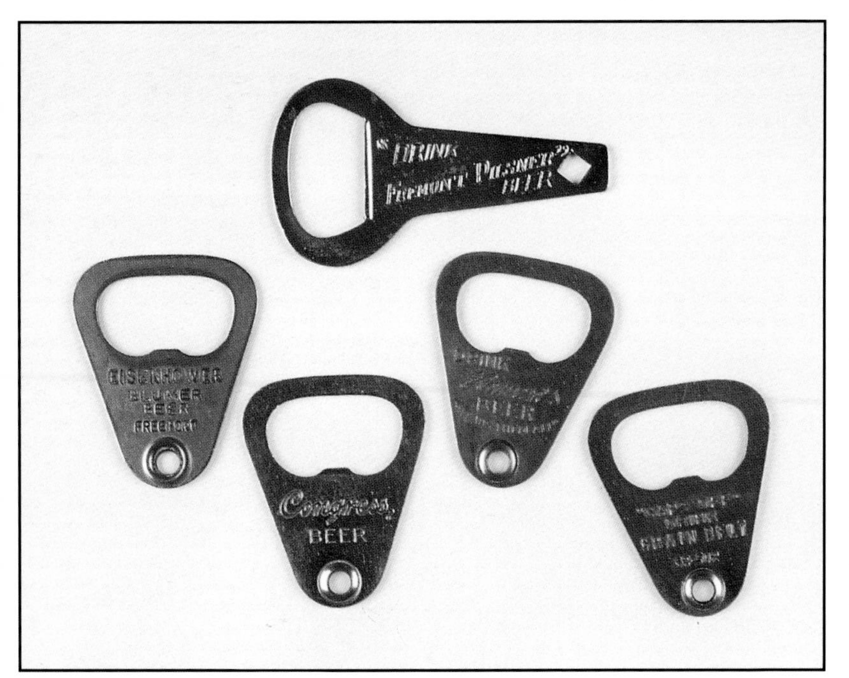

Top: C-3 Cap lifter with Prest-O-Lite key (diamond pattern). 2 3/4". $35-50.
Bottom four: C-8 Cap lifter. Key chain hole for use in key wallet. 2". $30-40.
One example says "Caps Off, Drink Grain Belt."

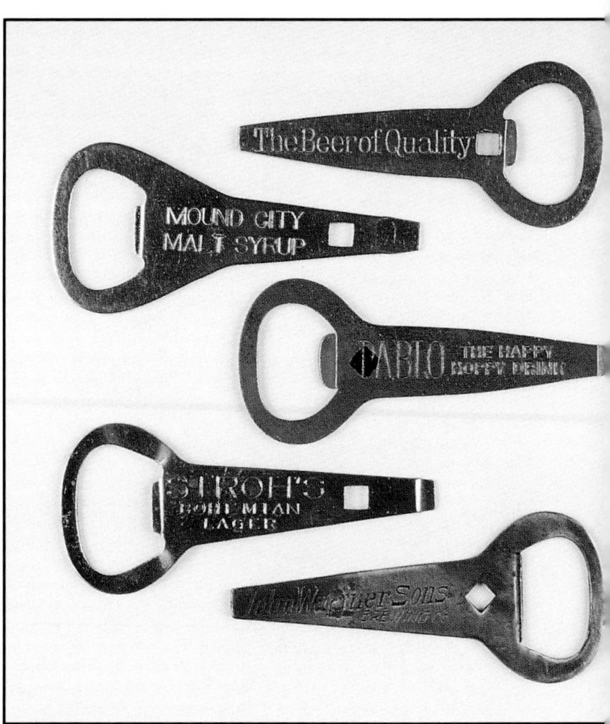

Top to bottom: C-4 Cap lifter with Prest-O-Lite key. 3 1/4". $5-10.
C-28 Cap lifter with Prest-O-Lite key. 3 1/4". $5-15.
C-29 Cap lifter with Prest-O-Lite key (Note: the diamond pattern is drawn in—original not located for photo). 3 1/4". $25-30.
C-37 Cap lifter with Prest-O-Lite key. 3 1/4". $25-30.
C-38 Cap lifter with Prest-O-Lite key (diamond pattern). 3 1/4". $25-30.

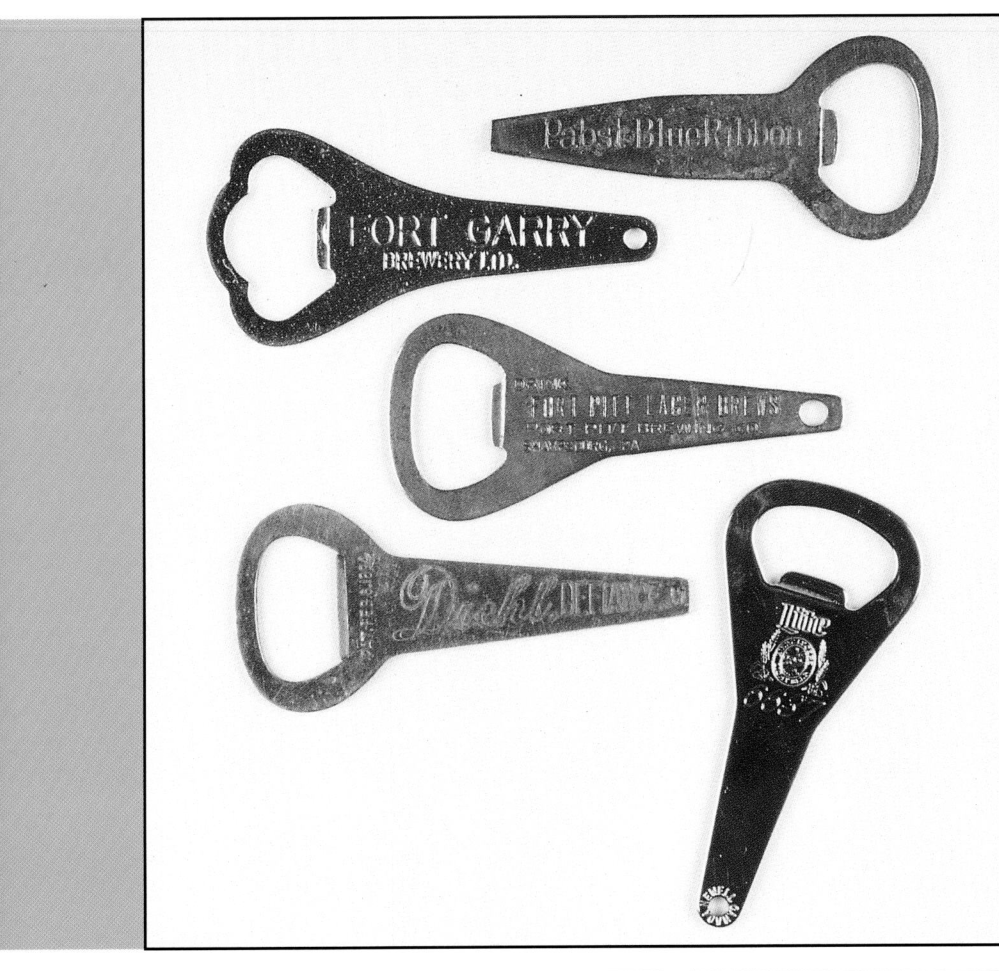

Top to bottom:
C-5 Cap lifter 3 1/4″. $5-10.
C-35 Cap lifter. 3 1/2″. $20-25.
C-36 Cap lifter. 3 1/4″. $10-20.
C-39 Cap lifter. Marked PAT. FEB. 6 1894. 3 1/4″. $25-30.
C-41 Cap lifter. 3 1/4″. $10-20.

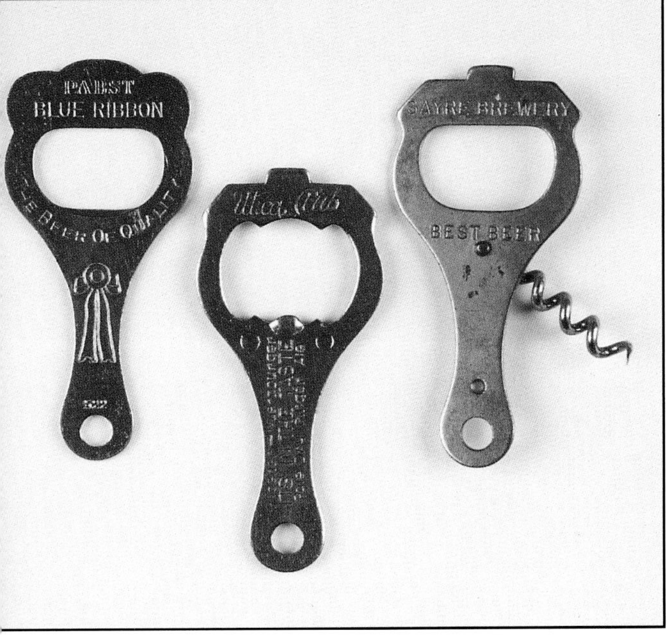

Left: C-6 Cap lifter. Rounded top. 3 1/8″. $2-10.
Middle: C-7 Cap lifter. Squared top. 3 1/4″. $2-10.
Right: C-26 Cap lifter with folding corkscrew. Marked MADE IN U.S.A. 3 1/4″. $40-50.

Top to bottom:C-9 Cap lifter. Tab in base. Handle 3/8″ wide at narrowest point. 3 1/8″. $2-20.
C-10 Cap lifter. Tab in base. Handle 1/2″ wide at narrowest point. 3 1/8″. $2-20.
C-11 Cap lifter. Tab above base. 3 1/8″. $2-20.
C-27 Cap lifter. Opener end bent 45 degrees. 2 7/8″. $15-20.

Left to right: C-32 Cap lifter. Tab in base. Straight sides and six-sided opener cut-out. 3 1/4". $15-20.
C-12 Cap lifter. Tab in base. 3 1/8". $1-30. *A souvenir opener for "Brockert's Ale" celebrates "DY Convention, Worcester, Massachusetts, 1936."*
C-13 Cap lifter. Tab above base. 3 1/8". $1-30.
C-14 Cap lifter. Tab above base. 3 1/4". $2-10.
C-15 Cap lifter. Tab in base. 3 3/8". $2-10.

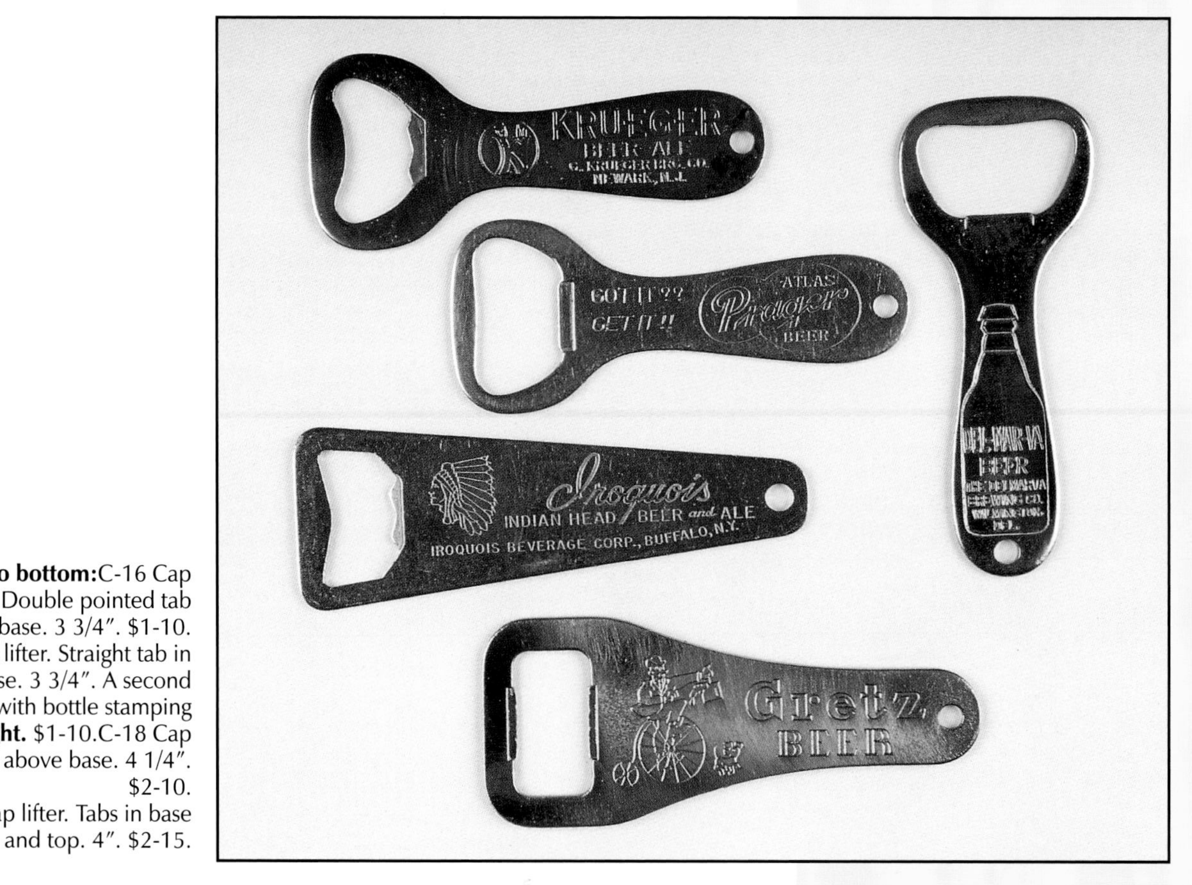

Top to bottom: C-16 Cap lifter. Double pointed tab above base. 3 3/4". $1-10.
C-17 Cap lifter. Straight tab in base. 3 3/4". A second example with bottle stamping is at **Right.** $1-10. C-18 Cap lifter. Tab above base. 4 1/4". $2-10.
C-19 Cap lifter. Tabs in base and top. 4". $2-15.

Top left: C-20 Cap lifter. Tab in base. Made by Veeder-Root. 4 1/2". $5-20.
Bottom left: C-21 Cap lifter. Tab in base. 4 1/4". $5-20.
Right: C-24 Cap lifter. Made by Veeder-Root. 4". Example shown $125-150; others $5-25.

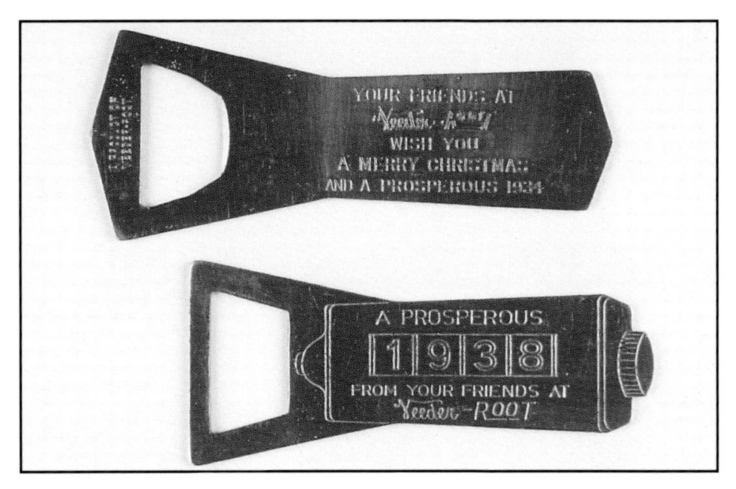

These openers were Christmas giveaways by Veeder-Root of Hartford and Bristol, Connecticut, in 1934 and 1938 with each one also having personal names stamped on the back.
Top: Type C-20 made by the Veeder-Root Co.
Bottom: A hybrid type C-24 also made by Veeder-Root Co.

Top to bottom: C-22 Cap lifter. Depicts hops, malt, and barley. Marked PAT NO. 91635. American design patent 91,635 issued to Ferdinand Neumer, February 27, 1934. 5". $3-5. *Advertises "Jacob Ruppert Brewer-New York, Save this opener, order by the case, Knickerbocker, the brew that satisfies."*
C-23 Cap lifter. Extra thick steel. 4 5/8". $15-20.
C-33 Cap lifter with screwdriver tip. 3 3/4". $15-20.
C-34 Cap lifter. Marked EKCO, CHICAGO. 3 1/2". $15-20.

Left: C-25 Cap lifter. Bottle shape. Marked METAL ARTS CO. MADE IN U.S.A. 3 1/2". $35-45.
Second left: C-40 Cap lifter. Bottle Shape. Marked VAUGHAN U.S.A. 3 1/2". $60-75.
Top right: C-30 Cap lifter with Prest-O-Lite key. American design patent 42,368 issued to Frank Mossberg, March 26, 1912. 3 1/8". $40-50.
Bottom right: C-31 Cap lifter with cigar box opener and nail puller. Marked M (in diamond logo) PAT.APPLD FOR. 3 1/8". $30-40.

Top left: C-42 Cap lifter. Guitar with mountains. 4 7/8". $5-7.
Bottom left: C-47 Cap lifter. Blimp. 5 3/8". $15-20.
Right: C-45 Cap lifter. Credit card size opener. May have magnet on back. Several color patterns. 3 3/8". $2-4. *Miller Brewing Company used this to advertise various brands with this advice: "Miller Brewing Company reminds you to please think when you drink." Miller Brewing Company had this type imprinted with the Miller logo for a 1991 National Sales Meeting in Palm Desert, California.*

Top left: C-43 Cap lifter. 3 1/2". $20-25.
Top middle: C-44 Cap lifter. 3 1/2". $15-20.
Middle: C-48 Cap lifter. 3 1/2". $3-4.
Right: C-50 Cap lifter. 2 1/2". $3-4.
Bottom: C-49 Cap lifter. Large key ring hole. 6 5/8" to 7 1/8". $3-4.

Cast Iron Cap Lifters

Cast iron openers were among the first cap lifters produced in the late 1800s when the crown cap was introduced. An opener similar to type D-6 appears in the drawings of Patent No. 514,200 issued to William Painter, Baltimore, Maryland, February 6, 1894. The patent claims "a capped bottle opener consisting of a suitable handle with a cap engaging lip adapted to underlie a portion of an applied bottle sealing cap, and also having a centering gage affording gaging contact with the side of the cap adjacent to the engaging lip, and still further affording fulcrum contact for enabling bearing engagement with the upper portion or top of the cap" (in other words—it removes a bottle cap!).

Top to bottom:(pair) D-1 Cap lifter with wire breaker. Made with raised and recessed letters. Curved cap lifter. 2 7/8". $3-20. D-5 Cap lifter. Recessed letters. Curved cap lifter. 3 5/8". $3-20. D-9 Cap lifter. Recessed letters. 3 1/8". $3-20. **(pair)** D-12 Cap lifter with loop seal remover. Recessed or raised letters. 3 3/8". $3-20. D-10 Cap lifter. Like D-12 but has gap at the top (missing loop seal remover). Recessed letters. 3". Not shown. $30-35.

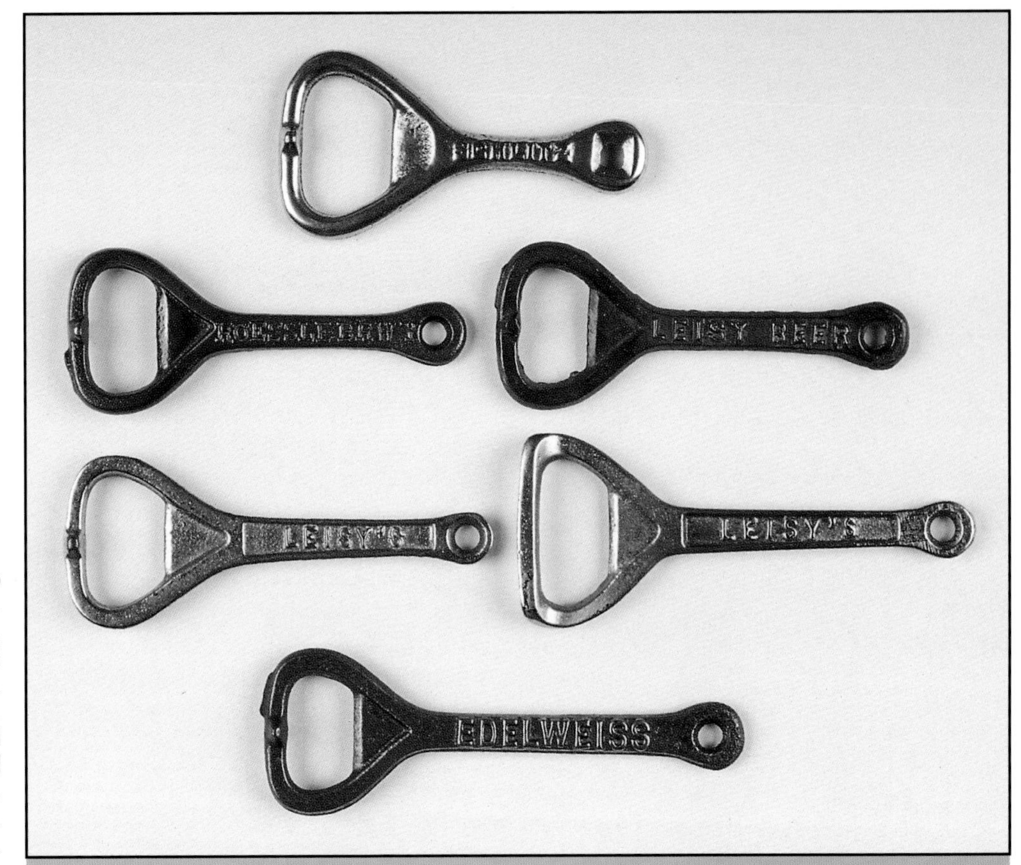

Top to bottom: D-2 Cap lifter with Prest-O-Lite key. Raised letters. 3". $3-20. **(pair)** D-3 Cap lifter. Raised and recessed letters. 3 3/8". $5-20. **(pair)** D-4 Cap lifter. Raised letters. Flat and curved opener ends. 3 3/4". $5-15. D-7 Cap lifter. Raised letters. 3 3/4". $30-40.

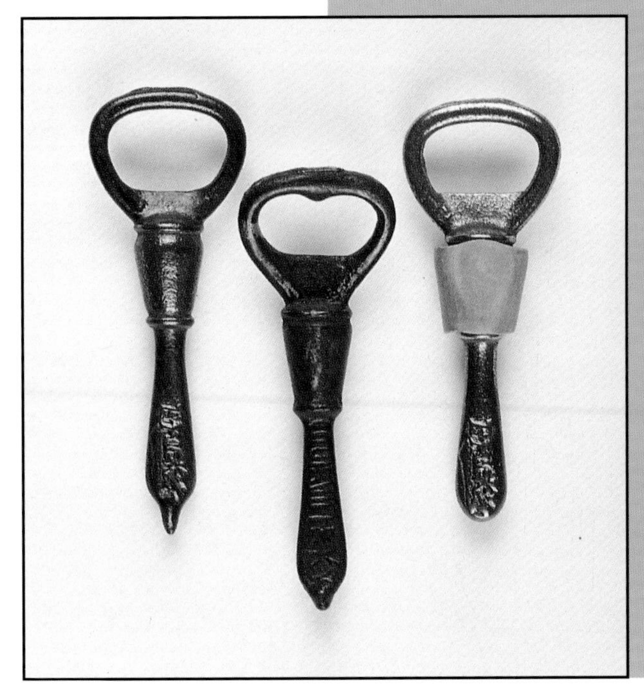

Left: D-24 Cap lifter with bottle stopper and aluminum stopper remover. Aluminum stopper remover at name end. 3 3/4". $30-40. **Middle:** D-21 Cap lifter with bottle stopper, loop seal remover, and aluminum stopper remover. 3 3/4". $40-50. **Right:** D-6 Cap lifter with bottle stopper and aluminum stopper remover. Aluminum stopper remover at opener end. Also made with screwdriver tip. American patent 514,200 issued to William Painter, February 6, 1894. 3 1/2". $15-20. *Found with advertising for the Frank Fehr Brewing Company of Louisville, Kentucky, with "Louisville" spelled with one "l" and with two "l's".*

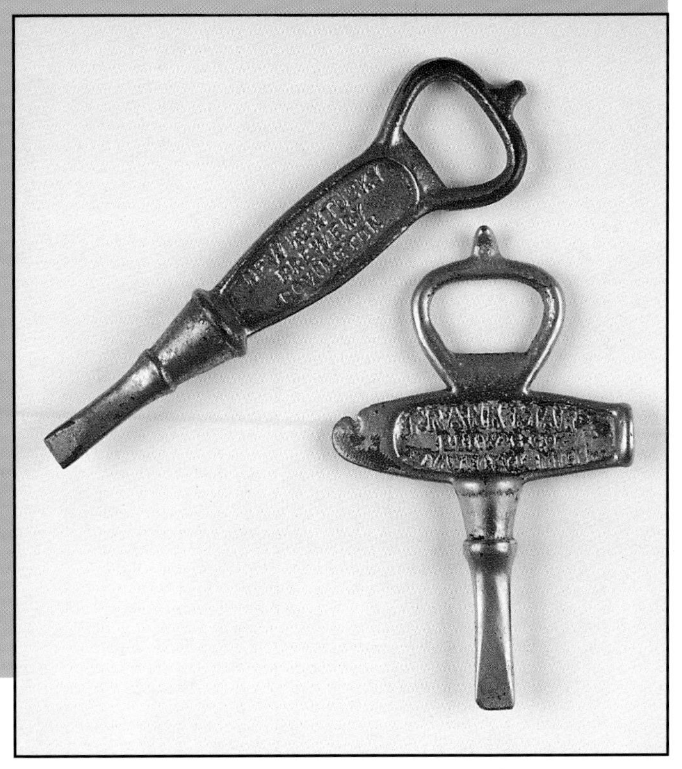

Top: D-16 Cap lifter with bottle stopper, loop seal remover, and screwdriver tip. 5 3/8". $125-150. **Bottom:** D-22 Cap lifter with horizontal handle. Includes bottle stopper, cigar box opener, nail puller, hammer, screwdriver tip, loop seal remover, and aluminum stopper remover. 4 1/8". $125-150.

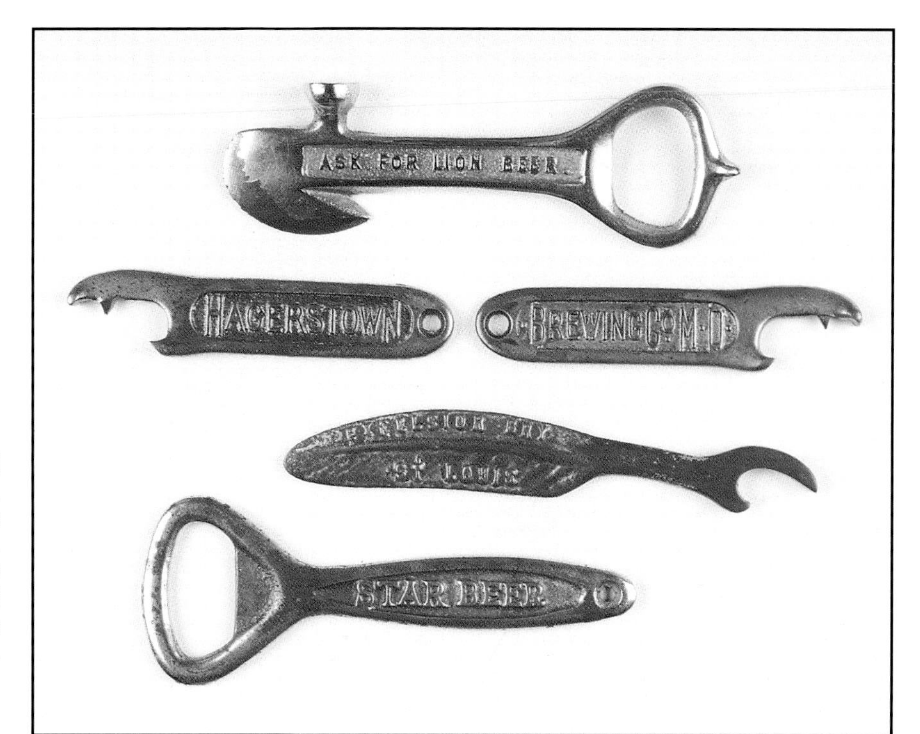

Top to bottom: D-8 Cap lifter with loop seal remover, cigar cutter, and tamper. 4 3/4". $75-$100. **(pair)** D-11 Cap lifter. Two points to grip cap. 3 3/8". $60-75. D-15 Cap lifter. Feather shape. 4 3/4". $125-150. *Only known with "Excelsior Bry. St. Louis, Red Feather Bottle Beer" advertising.* D-23 Cap lifter. 4 1/2". $40-50.

Left: **Left:** D-13 Cap lifter with circular handle that is a bottle resealer. 3". $40-50. **Middle:** D-14 Cap lifter with circular handle. 2 3/4". $15-20. **Right:** D-17 Cap lifter. 3". $5-8.

Right: Bottom of type D-13 showing bottle resealer.

Right: **Top right:** D-18 Cap lifter. Made expressly for Iron City Brewing Co. 3". $8-10. **Top left:** D-19 Cap lifter. Marked MADE IN U.S.A. Made expressly for Iron City Brewing Co. 3 1/2". $8-10. **Bottom:** D-20 Cap lifter. Marked COPYRIGHT AMINCO CO SOLID BRASS BOB. Made expressly for Iron City Brewing Co. 4 1/4". $8-10.

Below: Collector beware: This *Dixie Beer Co.* fantasy opener (Dixie Beer Co. never existed) was made about 10-15 years ago and at that time sold for $2-4. Present value is $5-8.

Wire Formed Openers

In 1915, renowned corkscrew inventor Edwin Walker of Erie, Pennsylvania, was granted an American patent (No. 1,150,083) for a wire formed cap lifter. The application for the patent was made in 1909 and documented it "as a new article of manufacture, a bottle cap lifter comprising a loop bent from a wire rod, and a plurality of lips swaged from the metal of the loop and extending inwardly from the inner contour of the loop." It was cheap to produce and, therefore, an extremely popular means of advertising a brewer's product. One manufacturer from Oakville, Connecticut, claimed, "We make the best and cheapest crown openers."

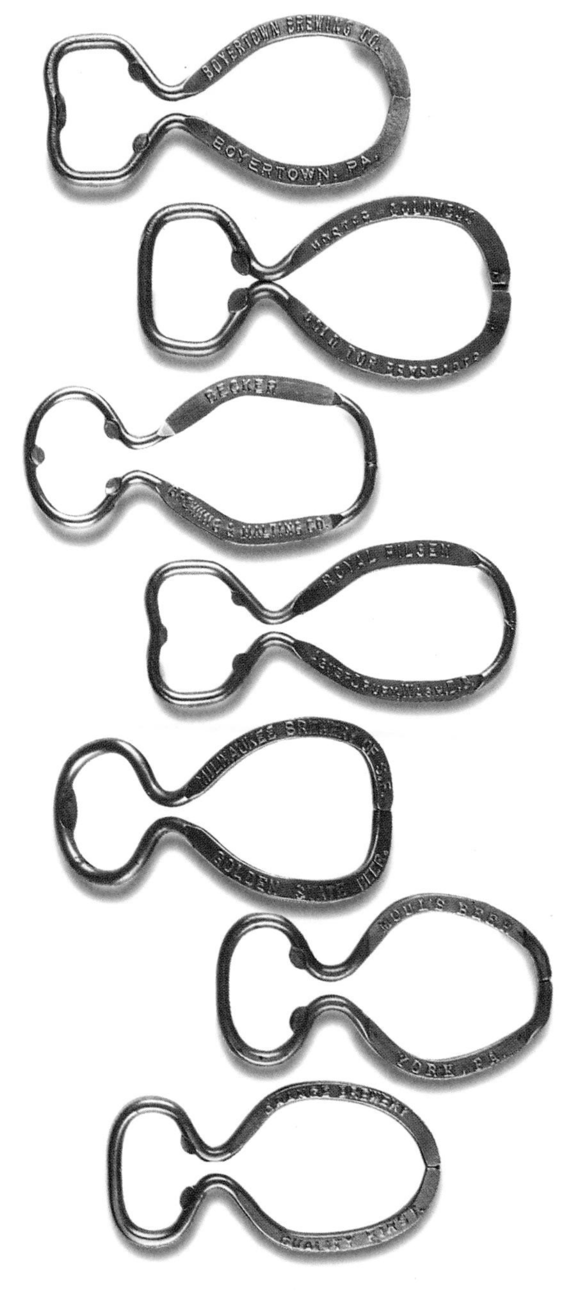

Top to bottom: E-1 Wide wire hoop. Three frets. Squared or dipped top. Flat handle at base. 3 1/2". $2-25. *Some brewers played up their "local" beer with advertisements such as "Daeufer's Beer, 'Allentown's' Favorite," and "In Jersey, It's Hensler's Beer & Ale."* E-18 Wide wire hoop. Squared top. Flat handle at base. Two frets. 3 1/2". $20-25. E-2 Wide wire hoop. Three frets. Rounded top. Rounded handle at base. 3 1/2". $2-25. E-3 Wide wire hoop. Three frets. Squared or dipped top. Rounded handle at base. 3 1/2". $2-25. E-22 Wide wire hoop bowed for better leverage. Rounded top. Flat handle. One fret. 3 1/4". $20-30. E-21 Wide wire hoop. Flat top. Rounded base. Two Frets. 3 1/4". $35-40. E-31 Wide wire hoop. Flat top. Flat base. Two frets. 3 3/8". $35-40.

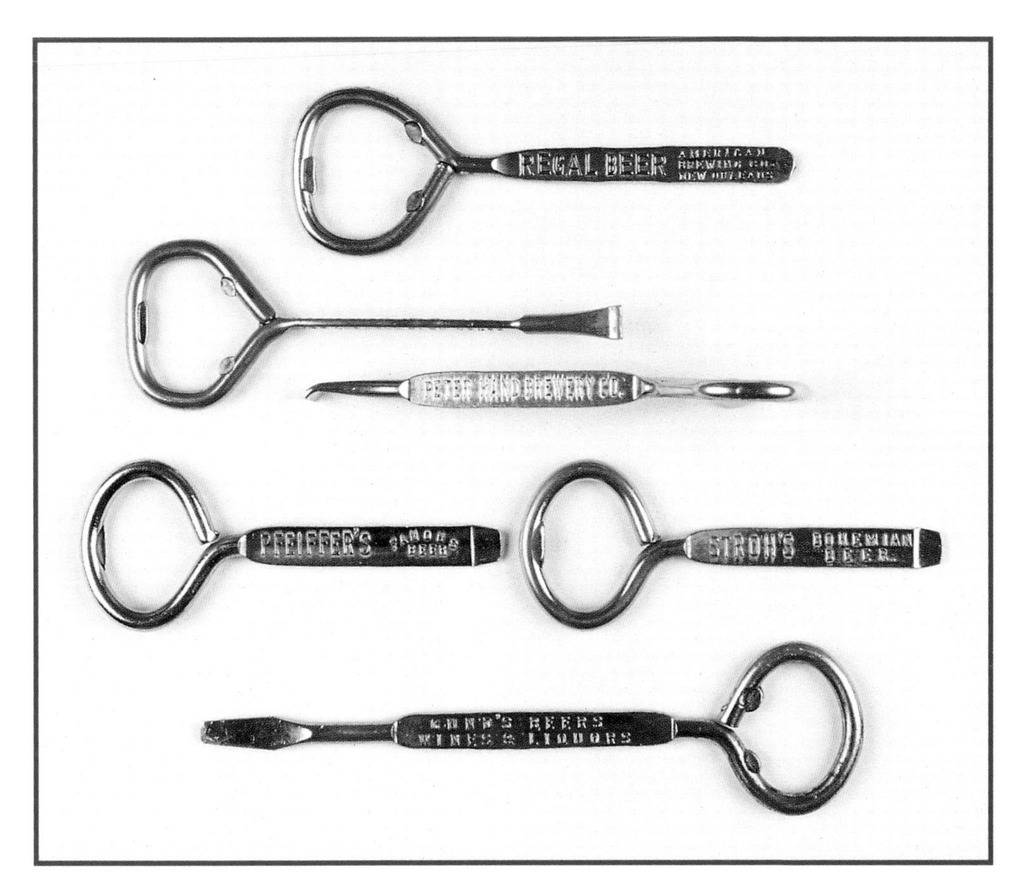

Top to bottom: E-4 Single wire handle. Some have screwdriver tips. One, two, or three frets. Rounded, squared, or dipped tops. 4 1/4" to 4 3/4". $2-40. **(pair)** E-15 Single wire loop with paint can opener handle. Squared top. 4 3/4". $10-30. **(pair)** E-27 Single wire loop. Thick handle. Rounded top. One or two frets. Some with screwdriver tip. 4". $20-25. E-28 Single wire loop with screwdriver end. Two frets. 6 1/4." $30-40.

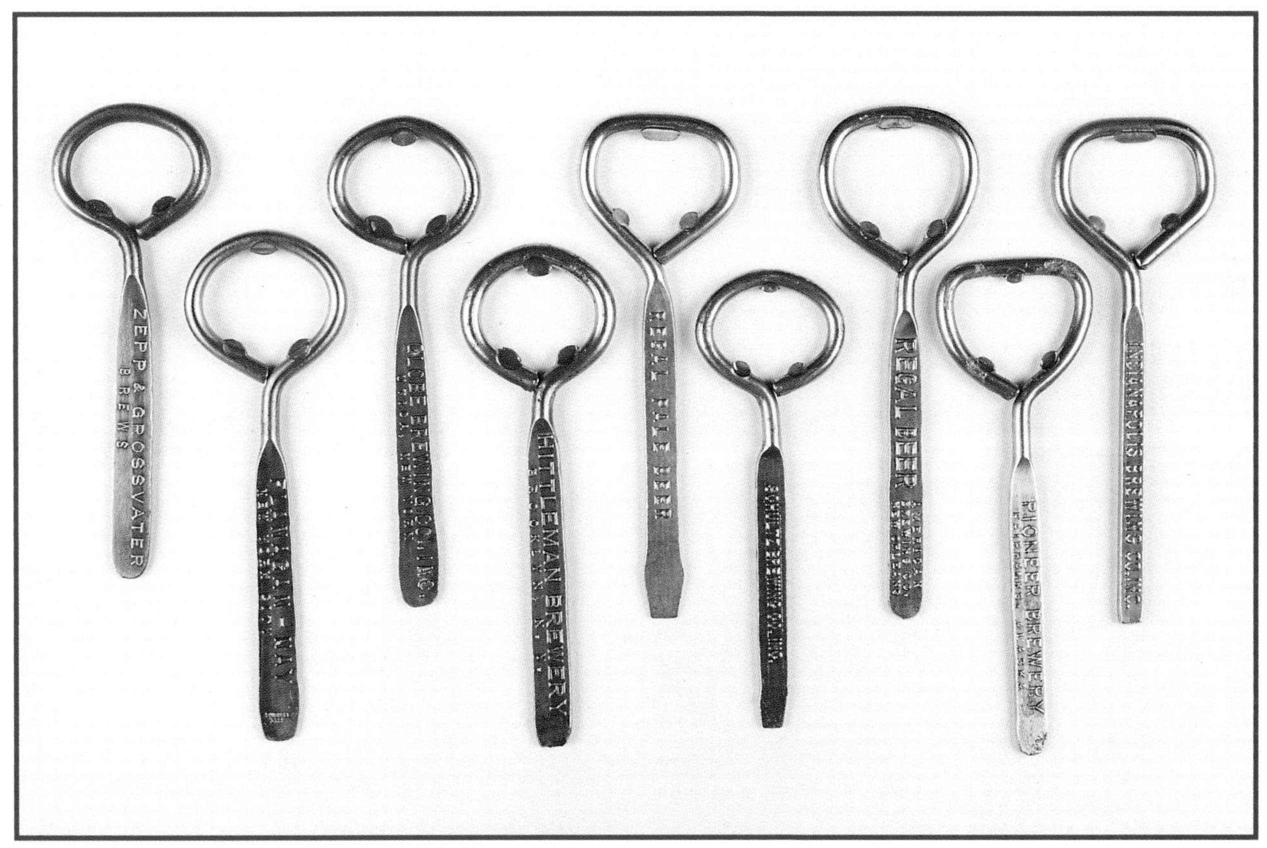

Variations of type E-4 include 2 and 3 fret openers, rounded and squared tops, heavy and light gauge wire, and flat end and screwdriver end openers. These variations could represent different types, but for the sake of keeping some sanity to the classification system, the various single wire handle openers are called E-4s. Types E-15 and E-27 are different enough to rate their own category. Over 300 different breweries and beer brands exist in type E-4.

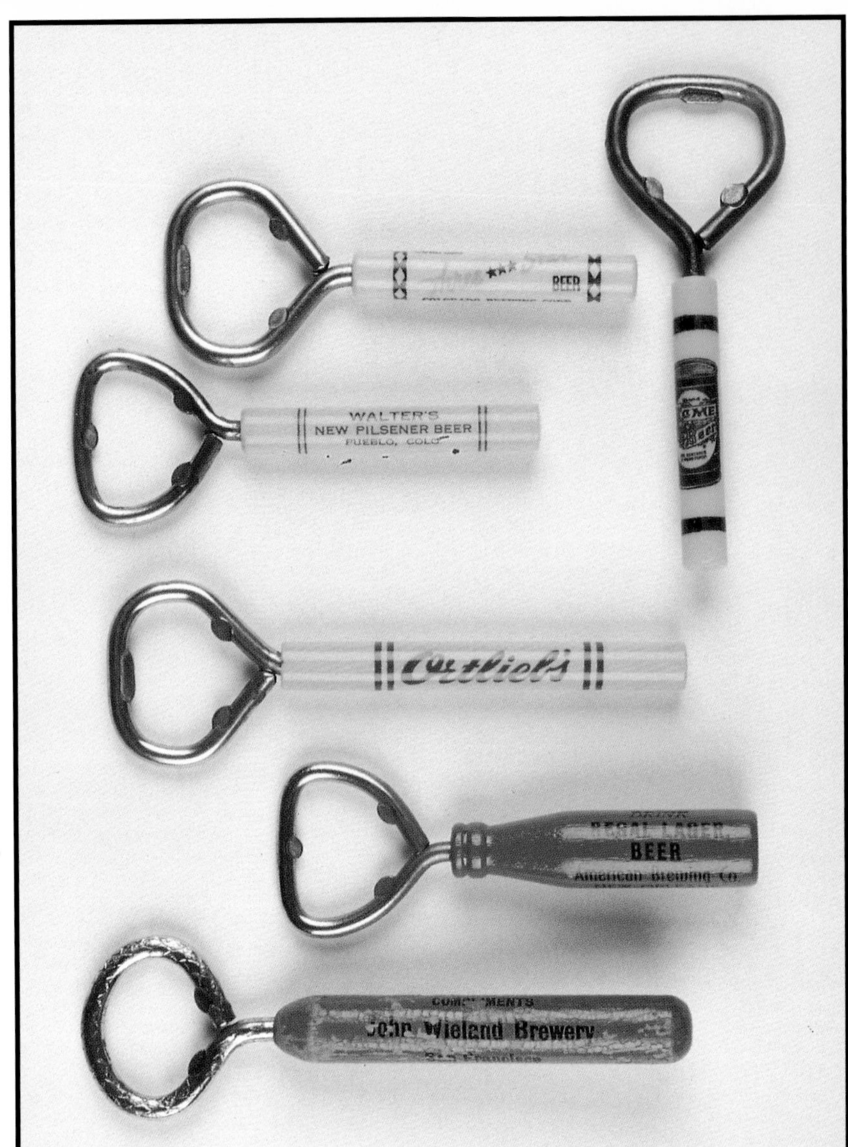

Right: **Top three:** E-5 Celluloid handle. Various colors. 4 1/4". $2-25. *One advertiser was stingy with the use of the opener: "To be used only with Drewrys Beer and Ale."*
Bottom three, top to bottom:
E-20 Celluloid handle. Various colors. 5 1/4". $5-10. E-19 Bottle shape wood handle with loop. Rounded or squared top. Two or three frets. 4". $20-30. E-24 Wood handle with wire loop. Two frets. 5 1/2". $40-50.

Below: E-19 variations including short fat bottles, tall skinny bottles, and sizes in between. Basically any wood bottle with a wire opener inserted is classified as an E-19.

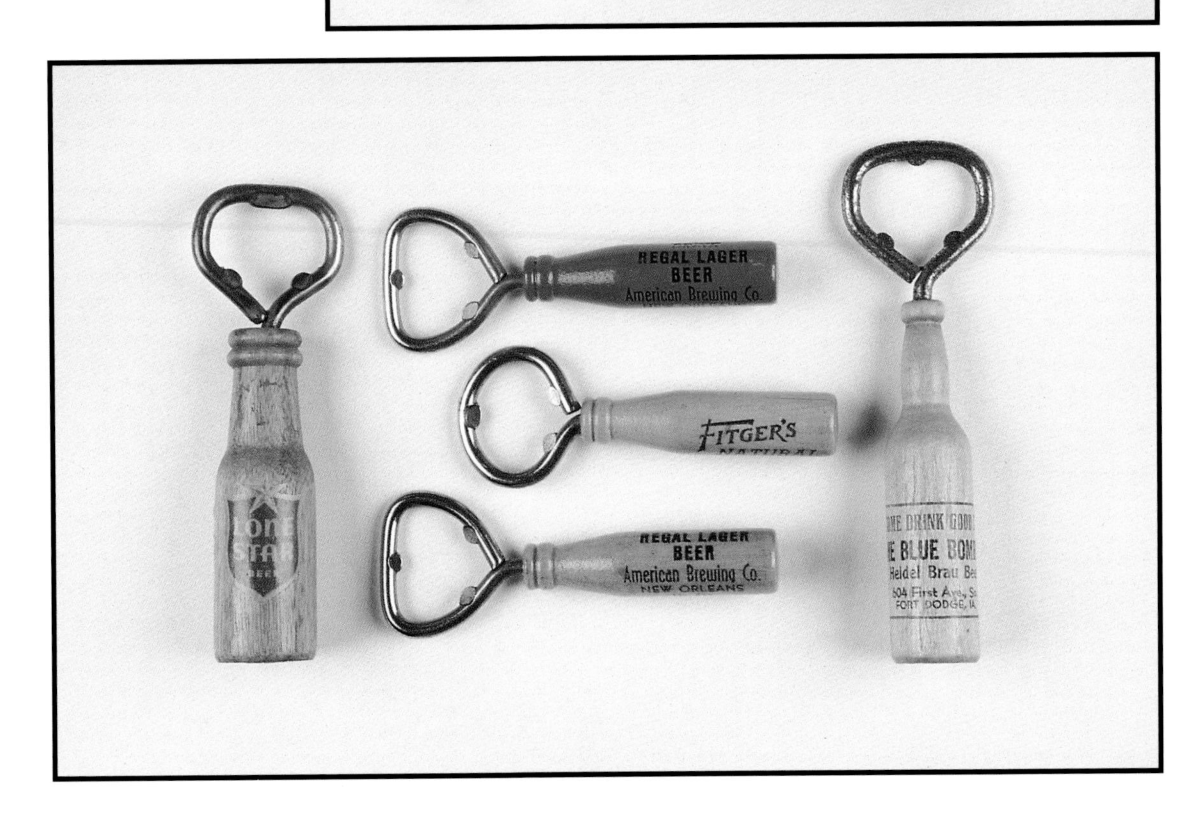

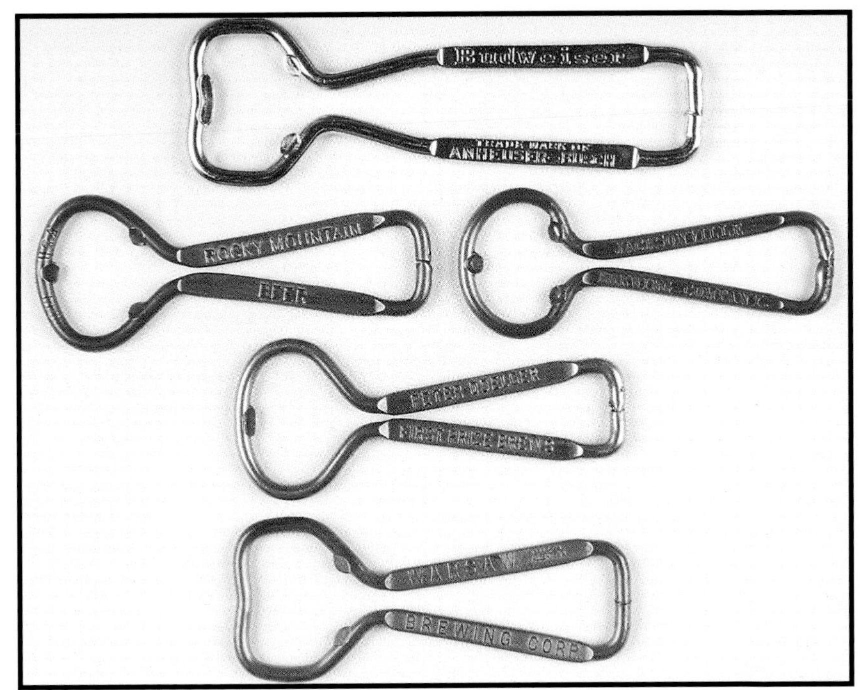

Top to bottom: E-11 Parallel stem handle. Squared or dipped top. Three frets. 4 5/8". $1-20. *The Brewers Association of America had this opener for their 1957 convention souvenir compliments of Emro Manufacturing of Chicago (see also I-14).* **(pair)** E-9 Single wire loop. Rounded top. Three frets. Some marked MADE IN U. S. A. 3 1/4" to 3 3/4". Two examples shown. $2-30. *In their advertising to sell the opener, Crown Cork and Seal Company, Baltimore, Maryland, stated: "You can have your name or other wording embossed on one or both sides and the finished product will be a much-sought-after souvenir highly appreciated by your trade. Its classiness and utility value coupled with your name will turn the trick."* E-29 Single wire loop. Rounded top. One fret. 3 1/2". $15-20. E-13 Single wire loop. Squared or dipped top. Two frets. 3 1/2" to 3 3/4". $2-25.

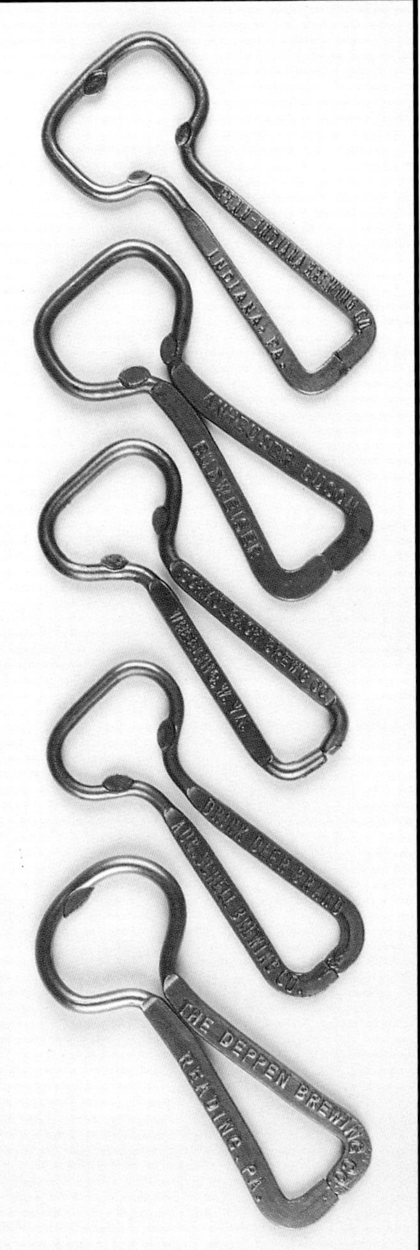

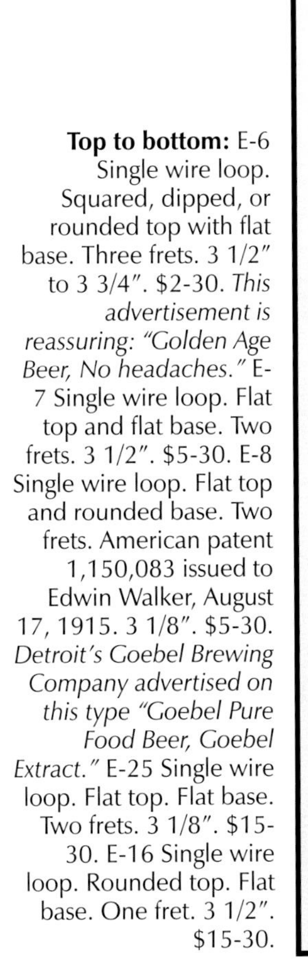

Top to bottom: E-6 Single wire loop. Squared, dipped, or rounded top with flat base. Three frets. 3 1/2" to 3 3/4". $2-30. *This advertisement is reassuring: "Golden Age Beer, No headaches."* E-7 Single wire loop. Flat top and flat base. Two frets. 3 1/2". $5-30. E-8 Single wire loop. Flat top and rounded base. Two frets. American patent 1,150,083 issued to Edwin Walker, August 17, 1915. 3 1/8". $5-30. *Detroit's Goebel Brewing Company advertised on this type "Goebel Pure Food Beer, Goebel Extract."* E-25 Single wire loop. Flat top. Flat base. Two frets. 3 1/8". $15-30. E-16 Single wire loop. Rounded top. Flat base. One fret. 3 1/2". $15-30.

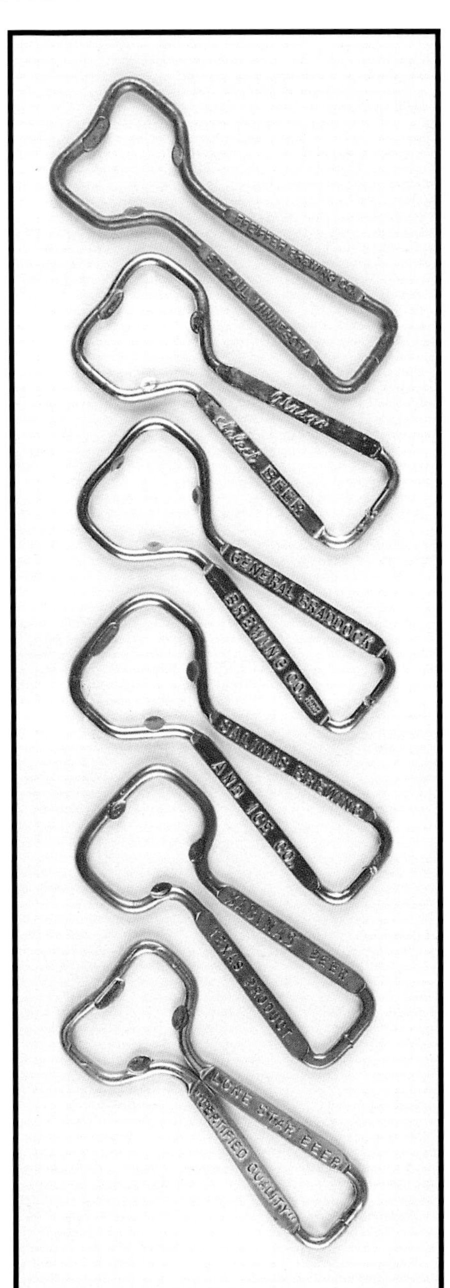

E-14 Single wire loop. Squared or dipped top. Three frets. 3 1/4" to 3 3/4". Six variations are shown $1-35. Over 800 different breweries and brands have been reported for this type. *Some of the great slogans on this type include: "The Pride of Rochester;" "Let's meet and be friends;" "The Thirst Choice of a nation;" "It's the Water;" "What'll you have?" Do you know the beers? Cataract Lager, New York; DuBois, Pennsylvania; Free State Brewery, Maryland; Olympia, Washington; and Pabst, Milwaukee.*

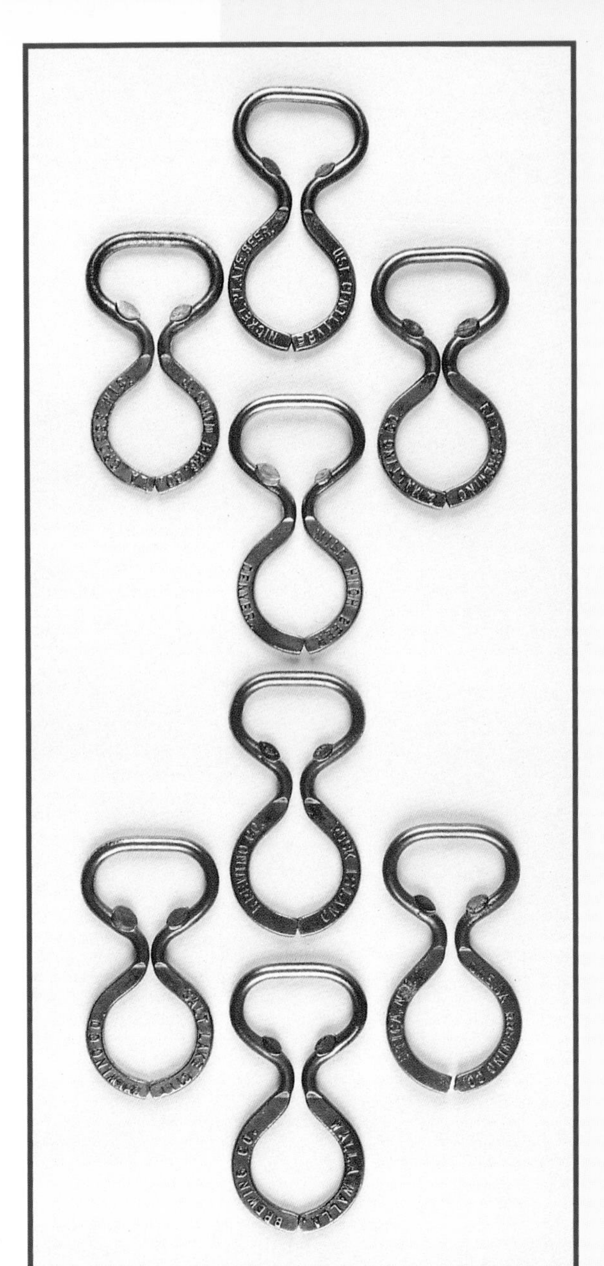

Right: E-17 Figure eight single wire loop. Flat top. Two frets. 2 1/2". $35-50.

Below: **Top three:** E-23 Single wire loop. Rounded top. Two frets. 3 1/4". $20-35.
Middle: E-30 Single wire loop with shield attached to base. Flat top. Two frets. 3 3/8". $40-50.
Bottom two: E-26 Single wire loop. Flat top. Two frets. 2 3/4". $40-50.

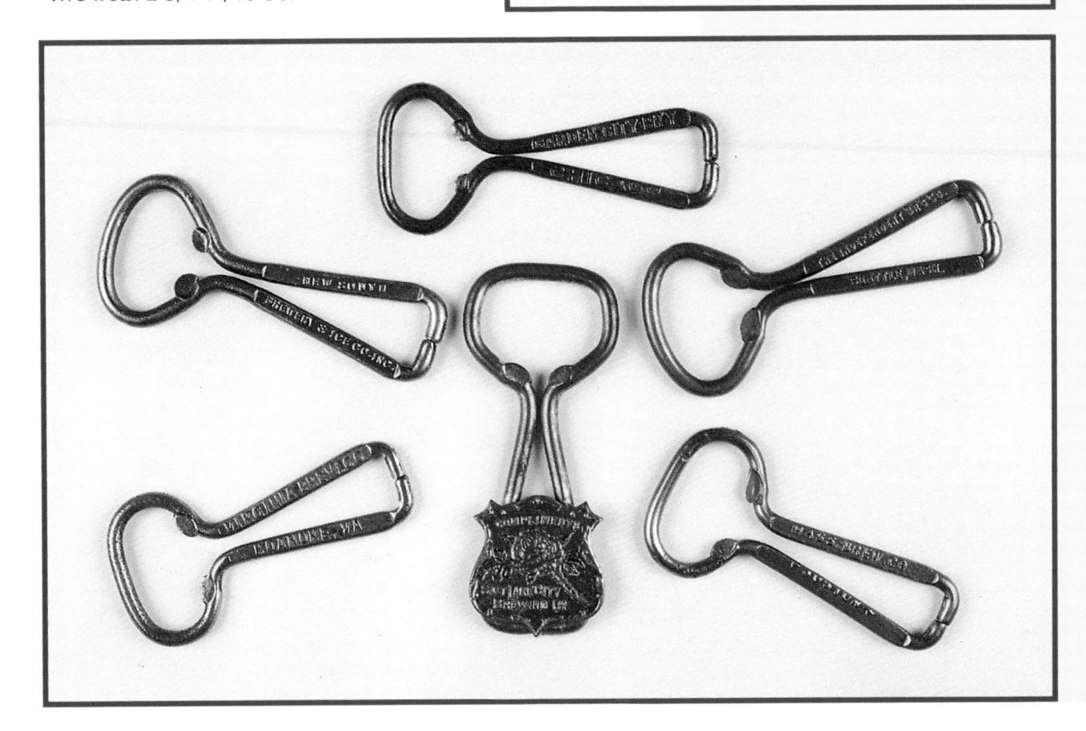

Multi-Purpose Openers

The advertising salesman who could sell a combination cake turner and bottle cap lifter to a brewery for advertising purposes deserves special recognition in the lore of openers. That unknown salesman evidently convinced both Becker Products in Utah and Gutsch Brewing Company in Wisconsin to buy this advertising gimmick. And he's probably the same guy that sold Cleveland & Sandusky Brewing Company on a cap lifter with spatula. Additionally, he found several brewers from Philadelphia, Pennsylvania, to Waterloo, Illinois, to purchase openers with slotted ladles.

What did the salesman have in mind for putting in beer when he sold a Wisconsin brewer a combination cap lifter and cocktail fork? At least the Nebraska brewer (Storz) that purchased the cap lifter with branding iron, could burn the Storz name into something (the mind of the buyer?). Do you like sardines and beer? A California distributor used a bottle opener and sardine can opener to advertise Wielands and Fredericksburg Beers. Were the cocktail spoons with beer advertising for mixing up a boilermaker? The ice picks seem to make the most sense—ready for a good cold iced bottle of beer?

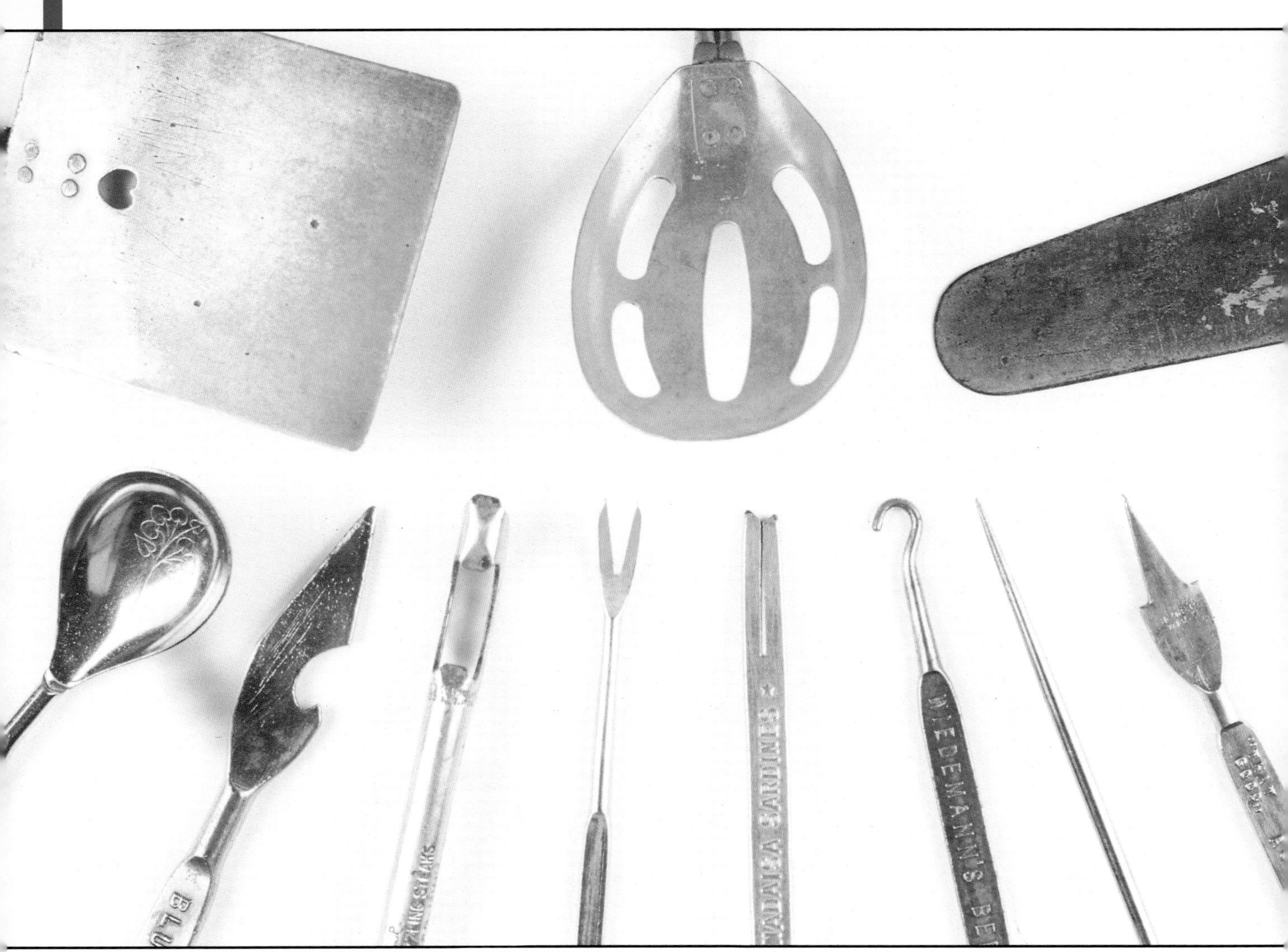

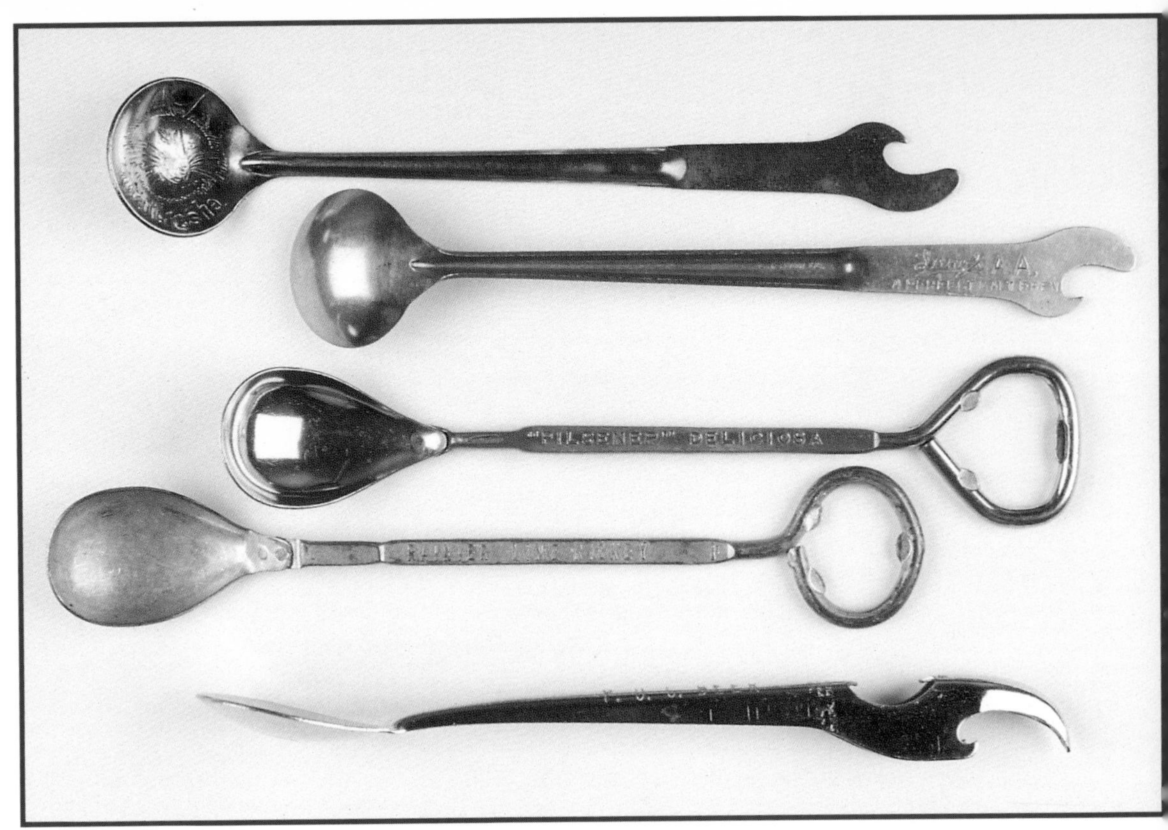

Top pair: F-1 Spoon with cap lifter. Made by L. F. Dow Company. 7 7/8". Two examples shown. $10-30.**Middle pair:** F-2 Spoon with cap lifter. American design patent 85,178 issued to Thomas Harding, September 22, 1931. The patent was assigned to the John L. Sommer Manufacturing Company. A 1937 Vaughan catalog says this is "the handiest combination of all for mixing drinks. Ideal for room service." 7 3/4". Two examples shown. $5-30. *Two advertising specialty firms put these messages on the spoons: "Bolms Bros., 354 4th Ave. N. Y. C., Advertising bottle openers & corkscrews" and "The Thompson Company, Dayton, Ohio, Advertising and merchandising."*
Bottom: F-3 Spoon with cap lifter and can piercer. American design patent 160,1950 issued to Le Emmette V. De Fee, October 31, 1950. Manufactured by Brown & Bigelow, St. Paul, Minnesota. Packaging for this names it "Spoonopener." 8". Side view only. $10-25.

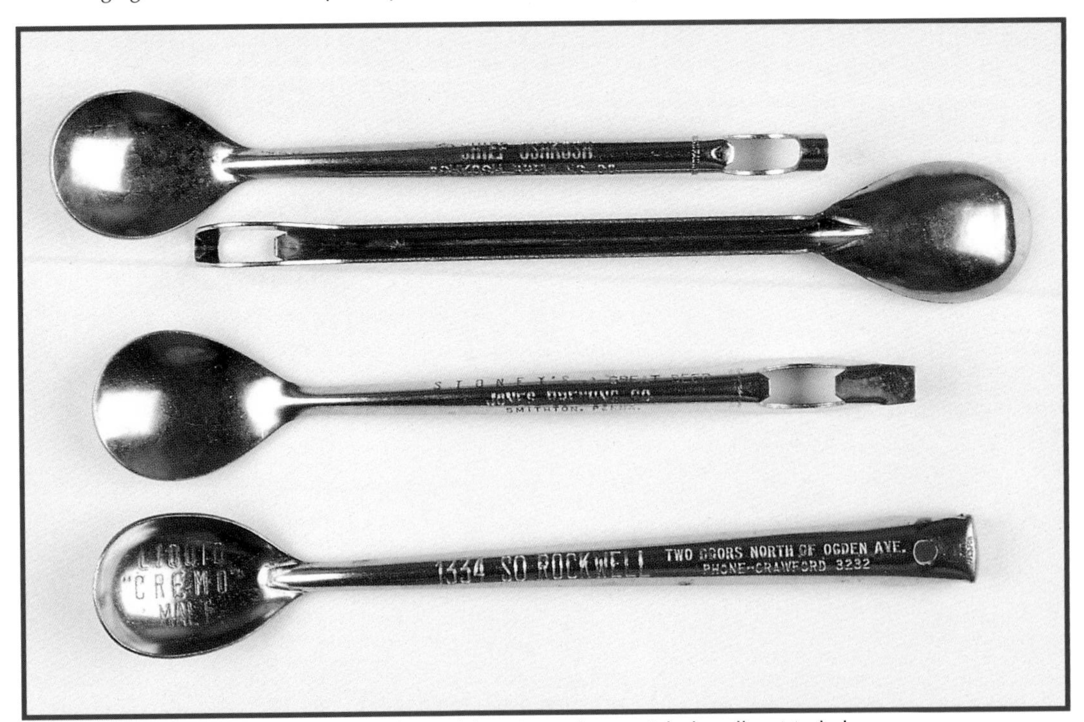

Top to bottom: (pair) F-4 Spoon with cap lifter. Curved or straight handles. Made by Brown & Bigelow. 7 1/4". $10-25. F-15 Spoon with cap lifter and screwdriver tip. 7 1/2". $10-25. *Fittingly Harry Young used this type for advertising "Why don't you stir up an acquaintance with Hamm's Beer & Carling's Ale?"* F-29 Spoon with over-the-top type cap lifter. Marked PAT APL FOR, BOTTLE OPENER. 8 1/8". $20-25.

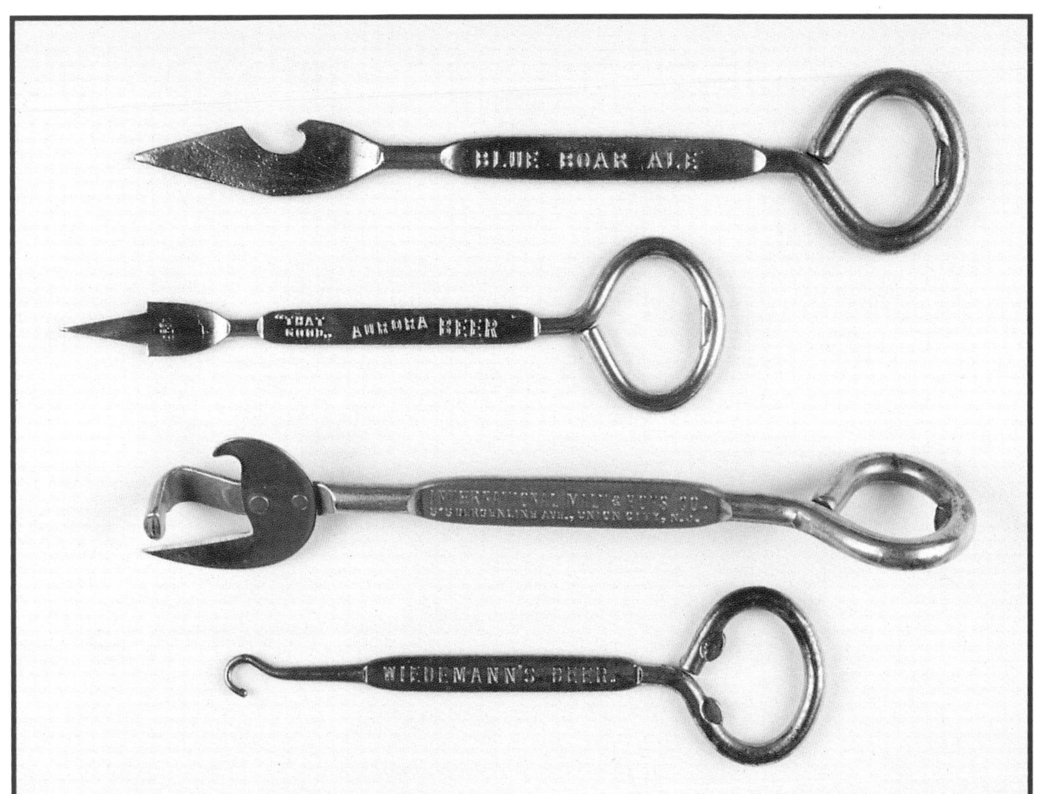

Top to bottom: F-5 Two cap lifters and ice pick. 7 3/4". $10-25. F-6 "Four in 1 Handy Tool." Bottle opener, friction cover opener, ice pick, and milk bottle cap lifter. One, two, or three frets. American design patent 43,278 issued to Thomas Harding, November 26, 1912. 6 1/4". $5-30. *One advertiser on this handy tool claimed "Altoona Brewing Co., The beer that builds you up." The Canton Brewing Company advertised its Topaz Lager Beer as "The jewel of them all," while Bohrer Brewing advertised Indiana Pride as "The gem of all bottled beer." Wayne Brewing Company, Erie, Pennsylvania, gave them out in a package that claimed only "3 tools in 1 - ice pick, crown opener & milk bottle cap remover."* F-26 Two cap lifters with tin can opener. 7 3/4". $30-40. F-30 Cap lifter and button hook. 6 1/8". $60-75.

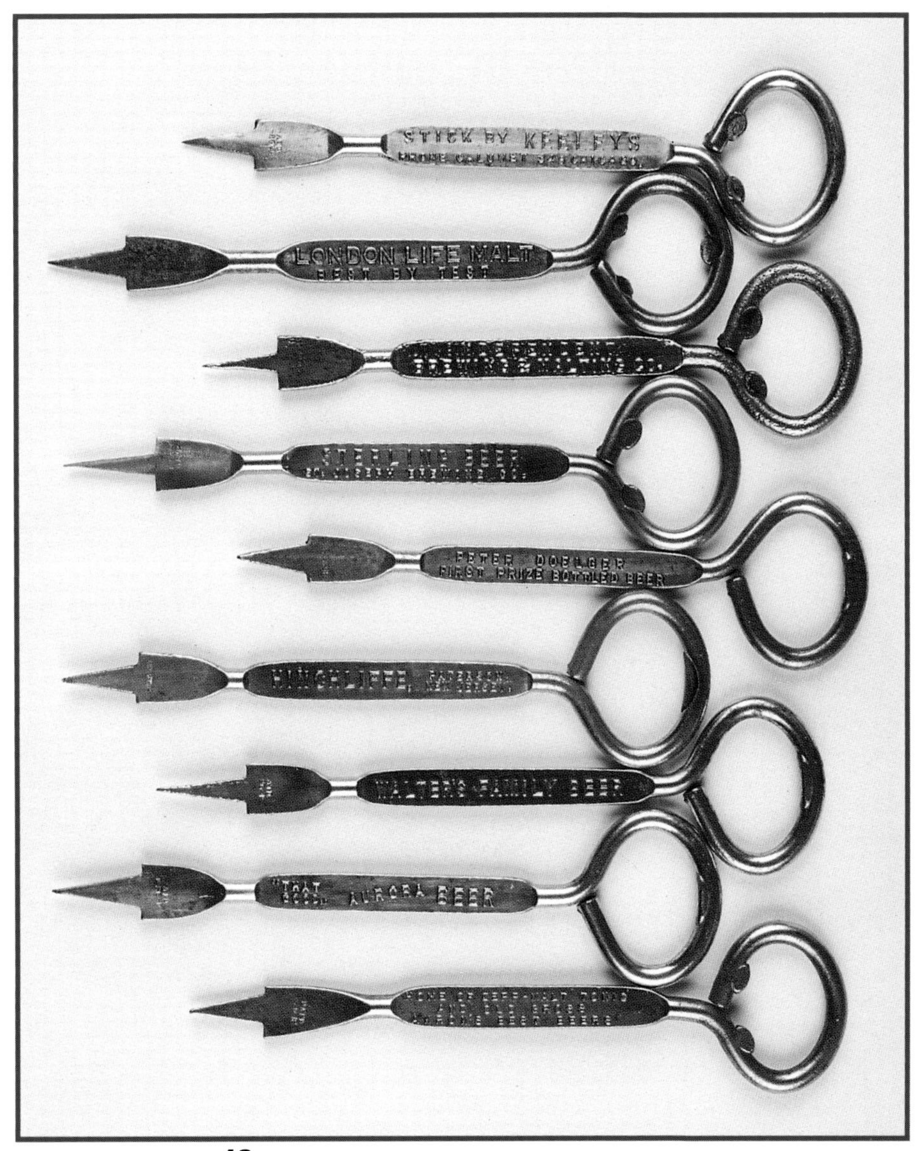

Like type E-4, type F-6 comes in several head types and with one, two, or three frets. Over 100 different breweries and beer brands are known to exist as advertising on type F-6. It is by far the largest number for a type in the multi-purpose openers category.

Top to bottom: F-7 Cap lifter with ice pick. Celluloid handle. Various colors. 7 1/4". $15-25. *Schick Distributing Company (Drewrys and Goetz) includes this advertising copy "At home or away, this can be used night and day."* F-8 Cap lifter with ice pick. Wood handle. 10". $10-20. F-24 Cap lifter with ice pick. Wood handle. 10 3/4". $30-40. F-19 Cap lifter with ice pick. Wood handle. 8 7/8". $30-40. F-25 Cap lifter with ice pick. Wood handle. 9 1/8". Cap lifter slides up pick to handle like F-19. $30-40.

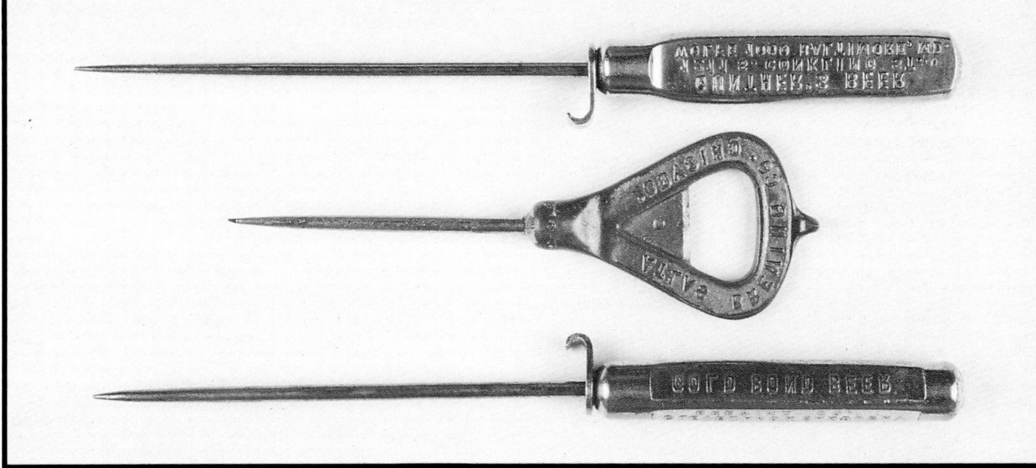

Top: F-9 Cap lifter with ice pick. Heavy (4 oz.) steel 4-Sided handle. 9 1/8". $30-60. **Middle:** F-14 Cap lifter with loop seal remover and ice pick. 5 3/4". $75-100. **Bottom:** F-32 Cap lifter with ice pick. Heavy (4 oz.) steel 2-Sided handle. 9 1/8". $50-60.

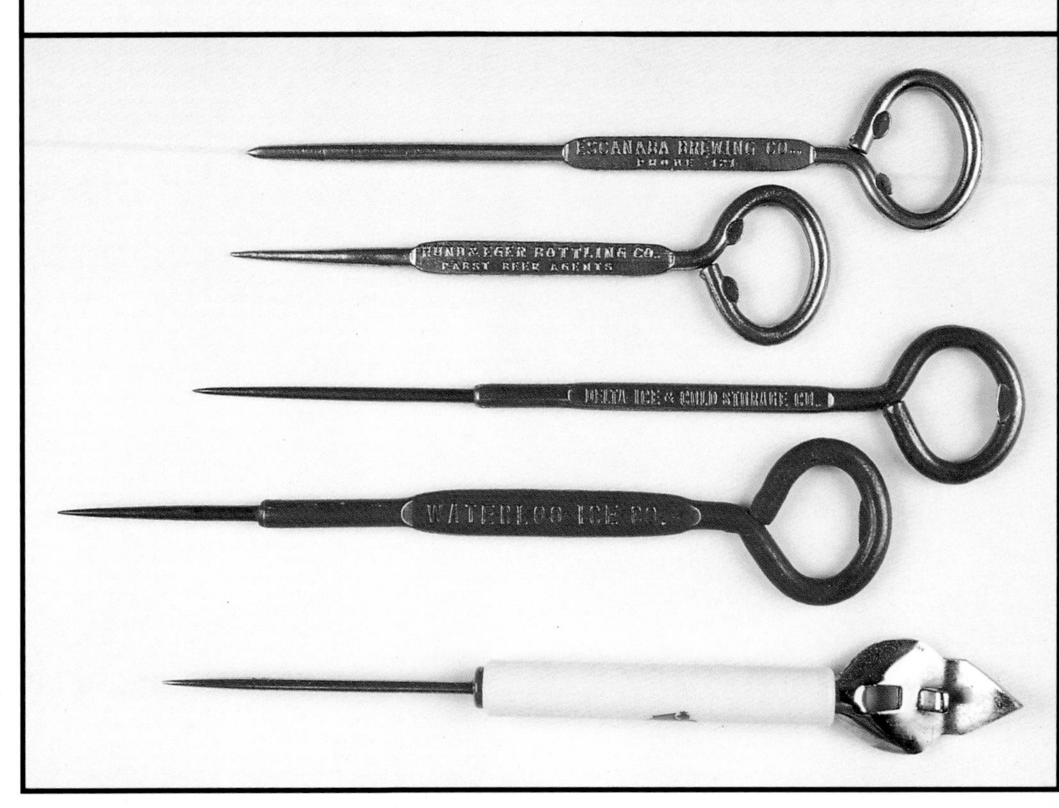

Top to bottom: (pair) F-18 Cap lifter with ice pick. One piece construction. American design patent 46,311 issued to Thomas Harding, August 25, 1914. 8 1/4". $25-40. **(pair)** F-27 Cap lifter with ice pick. Two piece construction. 8 1/2". $25-40. *The German Brewing Company, Cumberland, Maryland, advertised on this type "You choose well when you pick Old German Beer."* F-31 Cap lifter and can piercer with plastic handle. Ice pick end. Developed by Mr. Lipic of St. Louis, Missouri, in the early 50s. 8 1/2". $20-25.

Left: **Left:** F-11 Cap lifter and spatula. 10 1/2". $60-75. **Middle:** F-12 Cap lifter and slotted ladle. American design patent 47,016 issued to John L. Sommer, February 23, 1915. 10 1/2". $60-75. **Right:** F-13 Cap lifter and cake turner. American design patent 46,702 issued to John L. Sommer and Thomas Harding, November 24, 1914. 11 5/8". $60-75.

Below: **Left:** F-16 Cork puller. 6 1/4". $150-175. **Top right:** F-17 Cap lifter with cocktail fork. 8". $30-40. **Second right:** F-20 Cap lifter with branding iron. Made by Vaughan, Chicago. 6 7/8". $40-50. **Bottom right pair:** F-22 Cap lifter with sardine can opener. 6". $60-75.

Inset: Branding iron end of type F-20. Storz is the only known brewery advertising for this type.

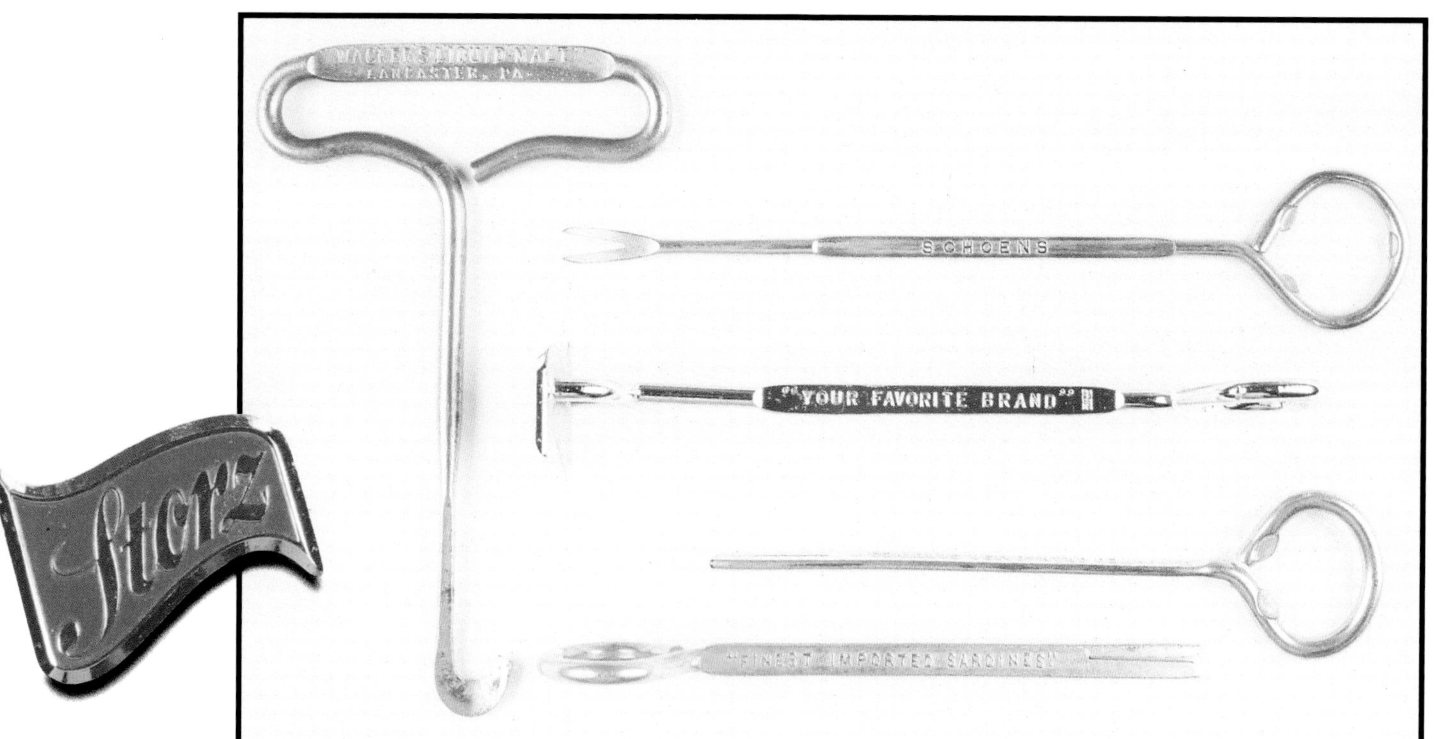

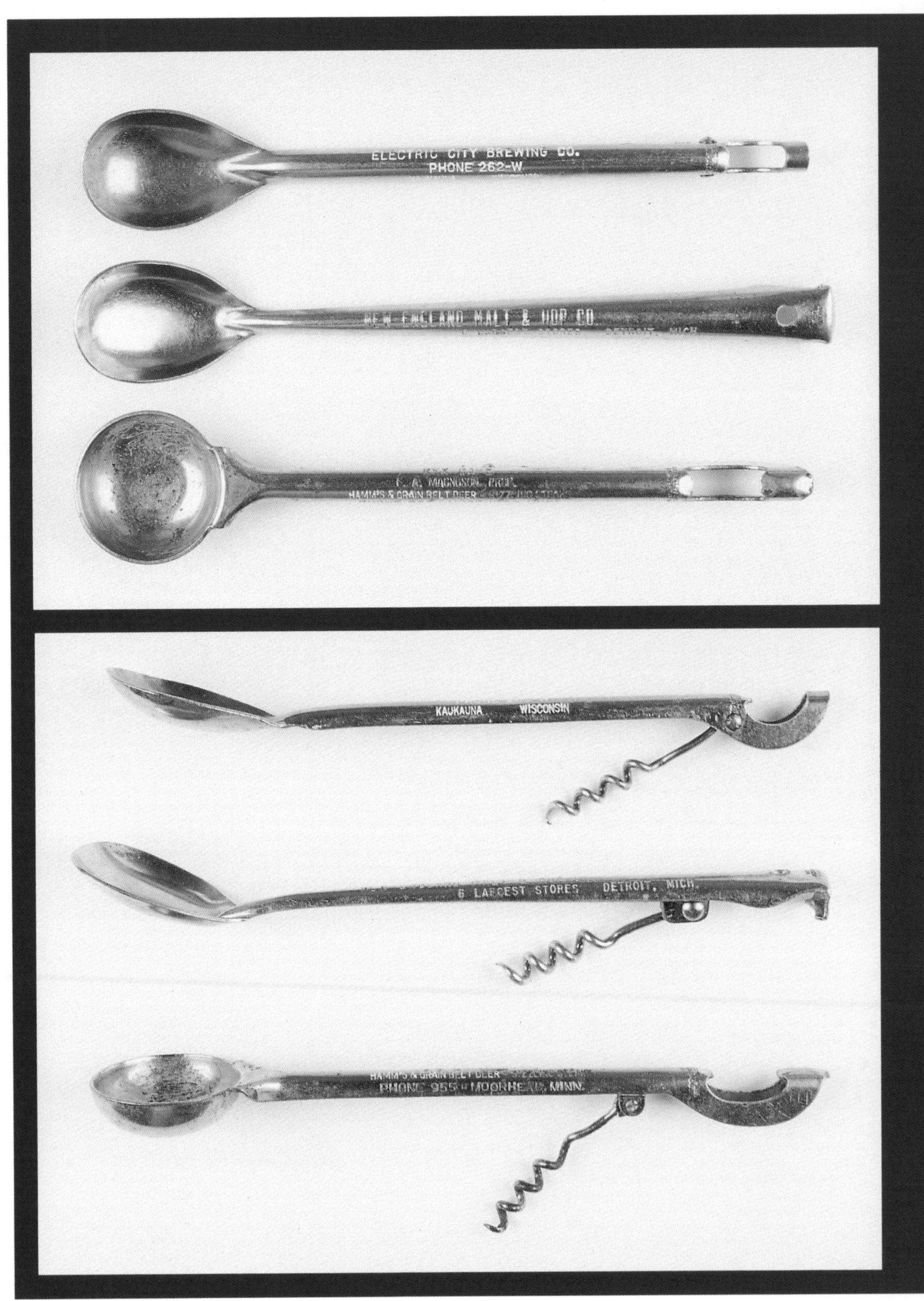

Top: F-21 Spoon with cap lifter and folding corkscrew. 7 1/2". $30-40.
Middle: F-23 Spoon with over-the-top type cap lifter and folding corkscrew. Marked PAT. APL'D FOR, BOTTLE OPENER. 8 5/8". $30-40.
Bottom: F-28 Spoon (deep round bowl) with cap lifter. 7 3/4". $30-40.

Formed Cap Lifters

This section contains cap lifters that have a three-dimensional look and feel. Some of the earliest cataloged examples were stampings with formed sides. The surfaces left plenty of room to carry the advertisers' messages. With a stretch of the imagination, one realizes that they can be easier to spot in a drawer and, by all means, easier to pick up than the flat types—ready for action. Harry Edlund patented the G-9 type in 1933 adding a wood handle and giving the advertisers a colorful opener for display of their message that was imprinted with a hot foil process. Later G types were cast with company logos and names and sold more often as souvenirs than as advertising openers.

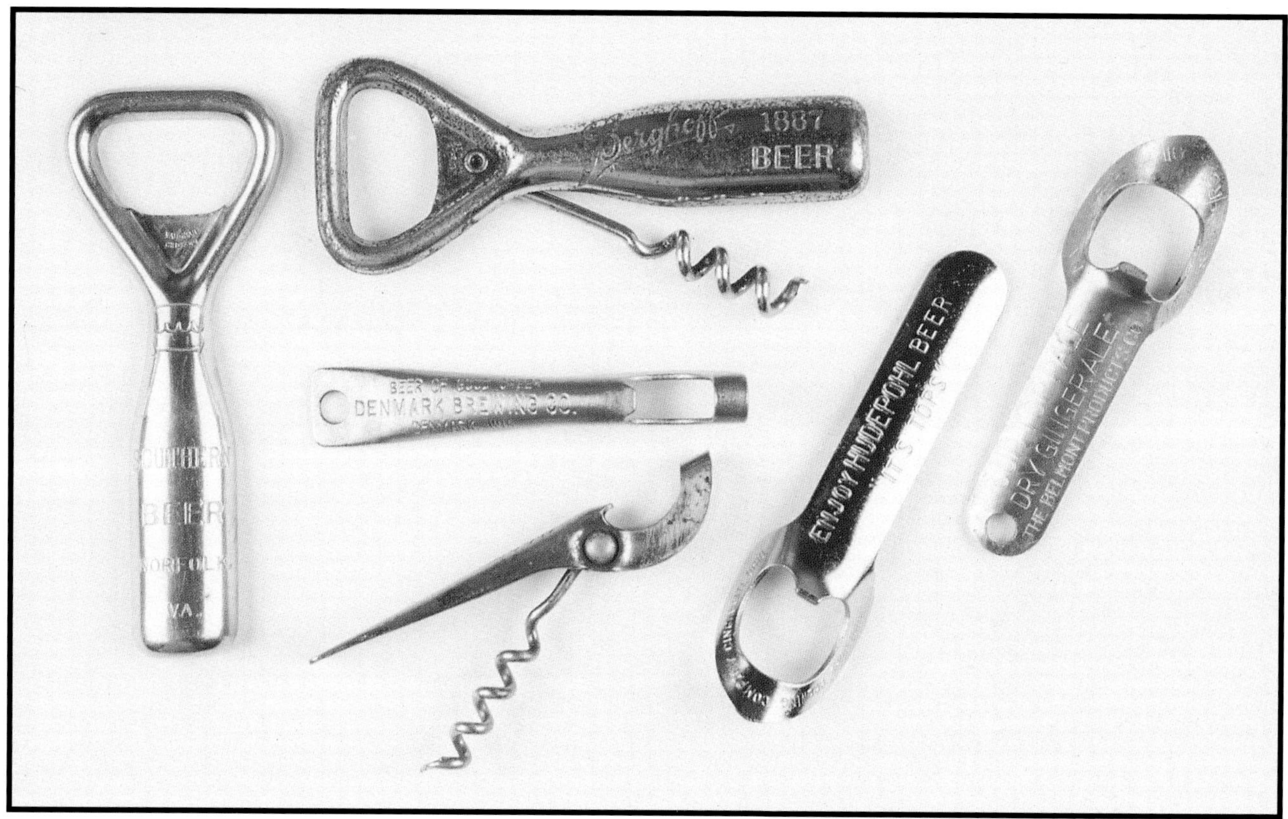

Left: G-1 Formed metal cap lifter. Bottle shape. Marked VAUGHAN CHICAGO. 4". $2-25. *Grand Prize used this type with two different claims: "South's Famous Beer" and "Texas' Largest Seller."* **Top middle:** G-40 Cap lifter. Formed metal. Bottle shape with folding corkscrew. Marked VAUGHAN CHICAGO. 4". $60-75. **Middle:** G-3 Cap lifter. Made by Brown & Bigelow (B & B) of St. Paul, Minnesota. 3 5/8". $2-15. *Who could refuse to heed Denmark Brewing Company's Wisconsin advertisement "Beer of good cheer?"* **Bottom middle:** G-53 Cap lifter with folding corkscrew. Made by Brown & Bigelow (B & B) of St. Paul, Minnesota. 3 5/8". $40-50. **Second from right:** G-2 Cap lifter with curved handle. 3 5/8". $2-15. **Right:** G-34 Cap lifter. Curved handle. 3 1/4". $10-12.

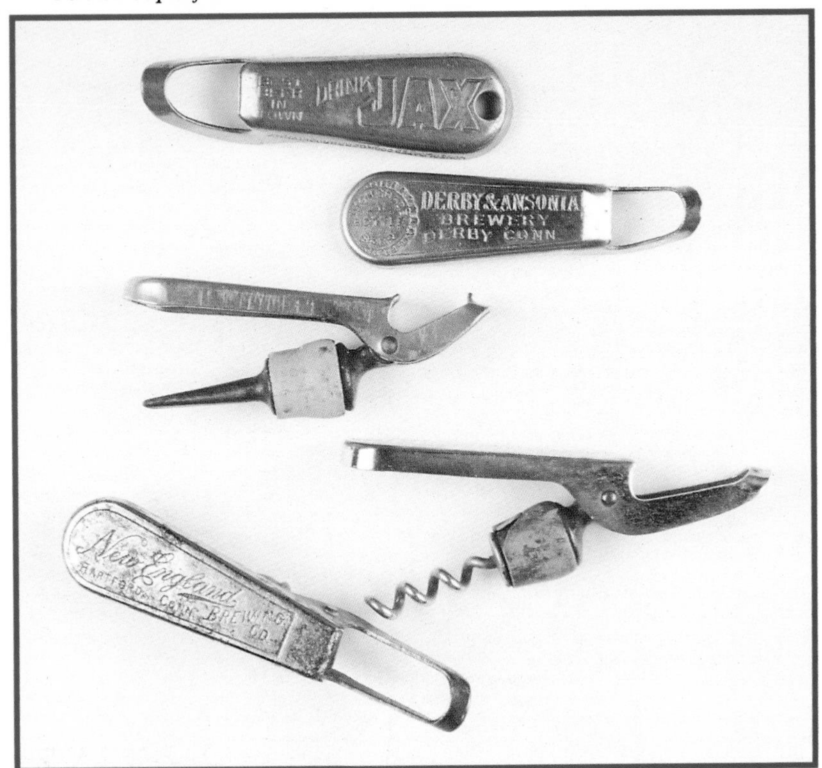

Top to bottom: G-4 Cap lifter with curved handle. On some these marks have been noted: CAPITOL STAMPINGS CORP. MILWAUKEE, WIS. and CONSOLIDATED CORK CORP. BROOKLYN N. Y. 3 1/4". $5-20. G-5 Cap lifter with flat handle. On some these marks have been noted: SEALTITE, 115 MAIDEN LANE, NEW YORK CITY PAT'D JULY 13, 1909, MF'D BY RYEDE SPECIALTY WORKS, PATENTS PENDING, MF'D BY RYEDE SPECIALTY WORKS, 137 MAIN ST., W., ROCHESTER, N. Y., PATENTED, and N.Y. SPECIALTY BY 'SESCO' PAT. JULY 13, 1909. American patent 928,156 issued to Adolph Rydquist, July 13, 1909. 3 1/4". $5-30. G-19 Cap lifter with folding bottle stopper. Marked MF'D BY RYEDE SPECIALTY WORKS, PATENTS PENDING. 3 1/4". $100-125. G-10 Cap lifter with folding corkscrew and bottle stopper combination. Marked MF'D BY RYEDE SPECIALTY WORKS, PATENTS PENDING. 4". $100-125. G-10 Top view.

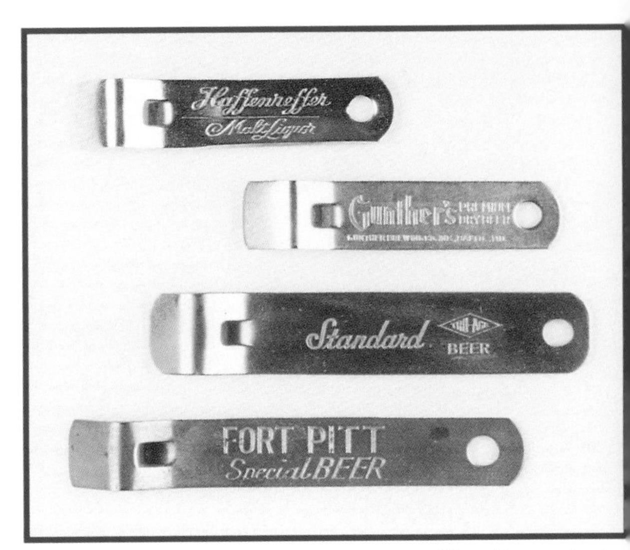

Top to bottom: G-6 Cap lifter with curved handle. Marked WALDEN, CAMBRIDGE, MASS. 3". $1-10. G-7 Cap lifter with flat handle. Marked WALDEN, CAMBRIDGE, MASS. 3". $1-10. *Walden used this type to advertise its own product with "Walden #103 catalog number, 16 Concord Lane, Cambridge 38, Mass."* G-8 Cap lifter with flat handle. Shorter and wider than G-22. Marked VAUGHAN MADE IN U.S.A. 4" to 4 1/8" long. 3/4" wide. $1-10. G-22 Cap lifter. Flat handle. Longer and narrower than G-8. 4 1/4" long. 5/8" wide. $15-20.

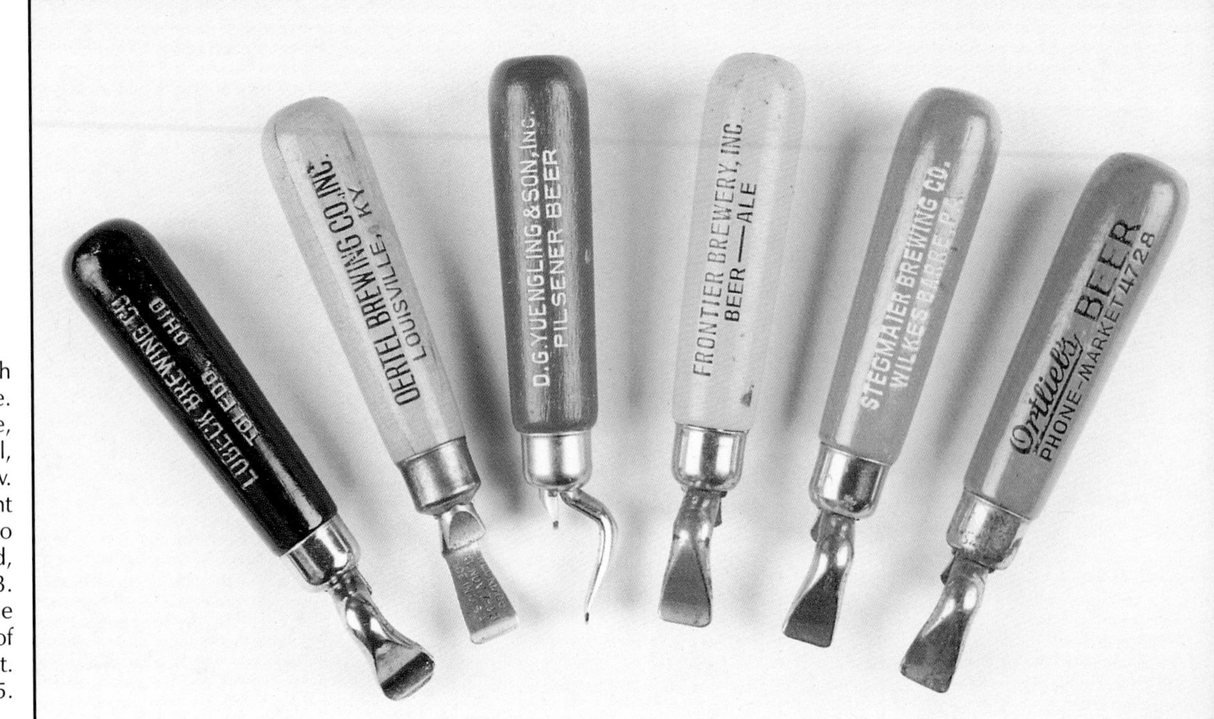

G-9 Cap lifter with wood handle. Colors: black, blue, gold, green, natural, red, and yellow. American patent 1,934,594 issued to Harry G. Edlund, November 7, 1933. Manufactured by the Edlund Company of Burlington, Vermont. 4 1/2". $2-25.

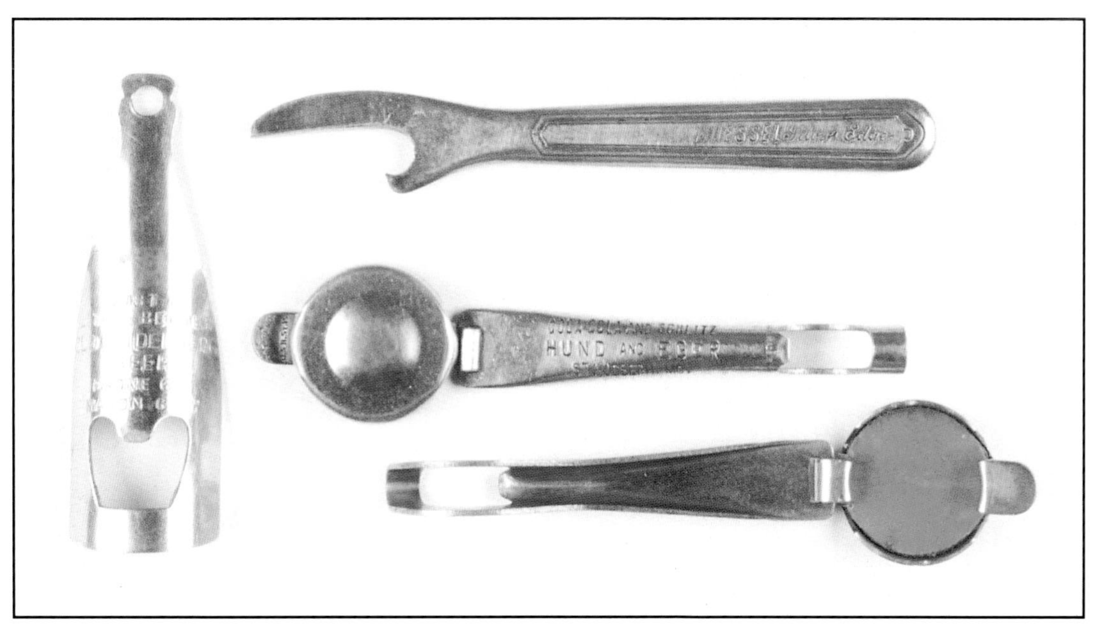

Left: G-21 Cap lifter. Curved bottle shape. 3 3/8". $15-25. **Right top:** G-11 Tableware cap lifter. 4 3/4". $75-100. **Bottom pair:** G-25 Cap lifter with recapper. Marked PAT.PEND. B&B ST.P. 4 5/8". $25-75. *A company representative handed out this type with "Compliments Schmidt's hired man in Minneapolis, Chas. Peterson."*

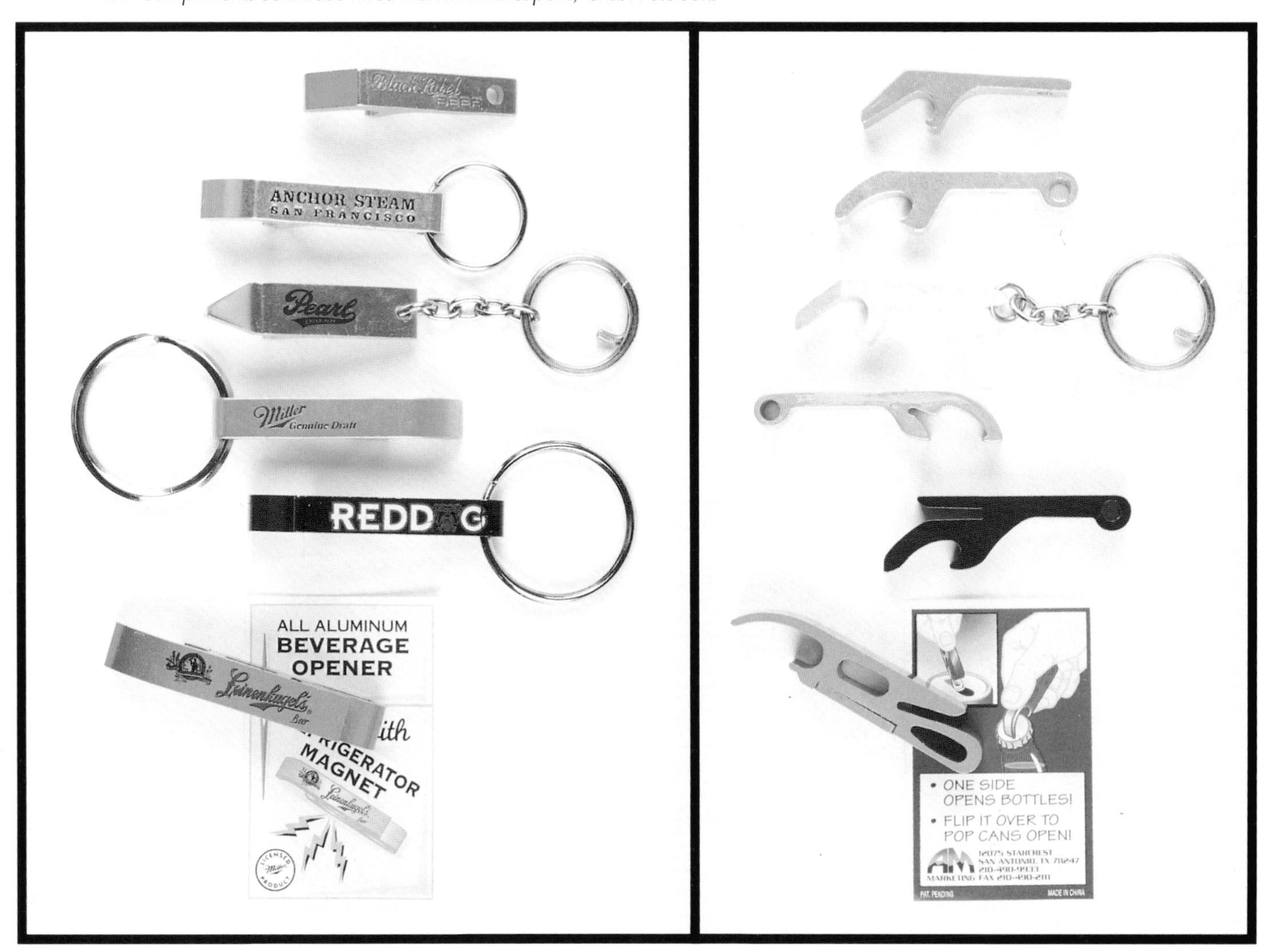

Top to bottom: G-12 Cap lifter. Lightweight metal. Marked CANADA. 1 7/8" $1-3. G-28 Cap lifter. Lightweight metal. Marked USA. 2 1/4". $1-3. G-45 Cap lifter/can piercer. Lightweight metal. 2". $3-5. G-46 Cap lifter/tab lifter. Marked USA PAT. 4,617,842 ROC PAT. 30850 MADE IN TAIWAN. 2 7/8". $3-5. G-48 Cap lifter/tab lifter. 2 5/8". $3-5. G-51 Cap lifter/tab lifter/refrigerator magnet. 2 1/2". $1-3.

Top: G-13 "Perfection" cap lifter. Marked VAUGHAN CHICAGO. 3 3/8". $3-15. **Second:** G-14 Cap lifter. Plastic advertising sleeve over metal. 3 1/8". $3-5. **Bottom left:** G-20 Cap lifter. Copper colored. Made by Century, Canada. 3 3/4". $8-10. **Photocopy:** G-50 Cap lifter. Copper colored. Made by Century, Canada. 3 3/4". $8-10. **Bottom right:** G-52 Cap lifter. Black with plastic insert. 2 1/2". $3-5.

Above:
Left top to bottom:
G-15 Cap lifter. 3 1/4". $5-8. *Coors.* G-16 Cap lifter. 3 1/2". $5-8. *Bud, The King of Beers.* G-18 Cap lifter. 3 1/2". $5-8. *Coors.* **Bottom right:** G-17 Cap lifter. 3 1/4". $5-8. *Oly Beer.*

Right:
Top: G-23 Cap lifter. 3 3/4". $5-8. *Olympia.* **Middle:** G-29 Cap lifter. 3 1/4". $5-8. *Coors.* **Bottom left:** G-24 Cap lifter. 3 1/2". $5-8. *Coors Light.* **Bottom right:** G-38 Cap lifter. 3 3/8". $5-8. *Coors.*

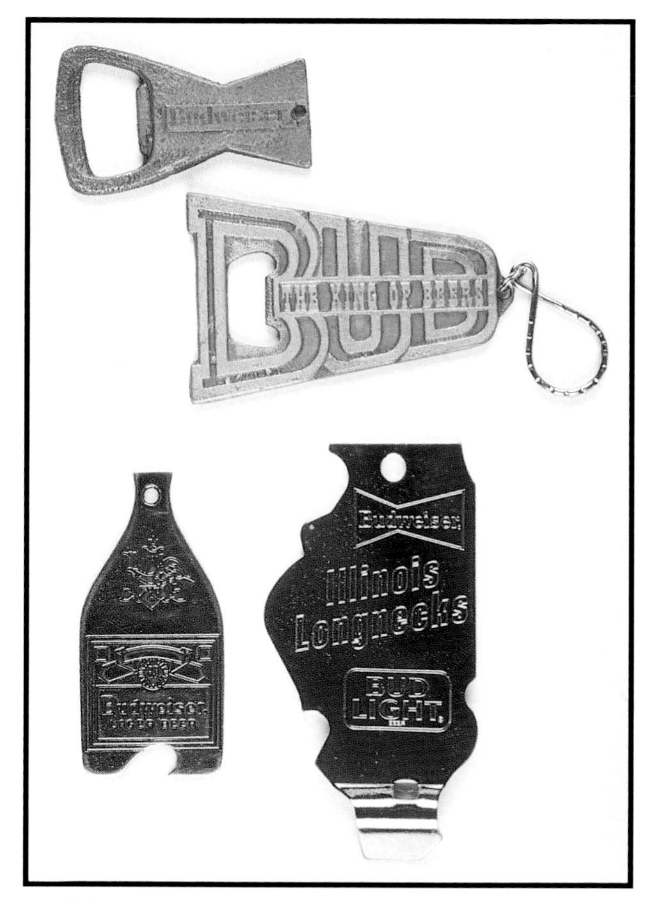

Top left: G-26 Cap lifter. 2 1/2". $20-25. **Top right:** G-30 Cap lifter. 3 1/4" $10-12. **Bottom left:** G-33 Cap lifter. Bottle shape. 3". $40-50. **Bottom right:** G-31 Cap lifter. State of Illinois shape. 4 1/8". $5-8.

Above: **Right side, top to bottom:** G-32 Cap lifter. Bottle shape. 7 1/2". $5-7. G-35 Cap lifter. Sheet metal stamping. 5 3/8". $3-5. G-42 Cap lifter. Brass with reinforcing ribs. 3 7/8". $5-7. G-43 Cap lifter. Brass with plastic insert. Marked COPYRIGHT AMINCO SOLID BRASS BOB. 4 1/8". $8-10. G-49 Cap lifter. Rectangular. Enameled brass. 3 13/16". $5-7. **Left side, top and bottom:** G-39 Cast iron cap lifter with plastic insert. 2 5/8". $3-5. G-44 Cap lifter. Solid brass. 3 1/16". $10-12.

Left to right: G-27 Cap lifter. Bottle shape. Marked MADE IN CANADA. 5". $8-10. G-36 Cap lifter. Bottle shape. 5 3/8". $3-5. G-37 Cast iron cap lifter with plastic insert. Bottle shape. 3 7/8" and 4 1/8". $3-5. **(pair)** G-41 Cap lifter. Bottle shape gray metal and enameled blue, gold, green, and white. 4". $8-10. G-47 Cap lifter. Bottle shape. Plastic covered steel. 6 3/8". $5-7.

Over-the-Top Type Cap Lifters

In 1924, Harry L. Vaughan of Chicago, Illinois, was granted Patent No. 1,490,149 for the invention of the over-the-top type opener. The patent application was filed in 1921 and was assigned to the Vaughan Novelty Manufacturing Company. The cap lifter goes over-the-top of the cap and fits under the far edge of the crown. In a 1922 Vaughan catalog, the opener is presented with these comments: "A slight downward pressure and off comes the cap in the hand. Everybody says—The best Bottle Opener ever invented." Prices ranged from $7.50 for 250 to $17.00 per thousand in 10,000 quantity. Vaughan advertised this type with one stamped "Over the top bottle opener manufactured by Vaughan Novelty Mfg. Co., Inc., Chicago, Illinois, Trade Mark, Made in U. S. A."

The Mergott Company of Newark, New Jersey, imitated the over-the-top type with a folding corkscrew added on the underside per William Hiering's 1928 and 1929 patents. This was followed by the production of a number of colorful bottle shape over-the-top versions lithographed by the Muth Company in Buffalo, New York.

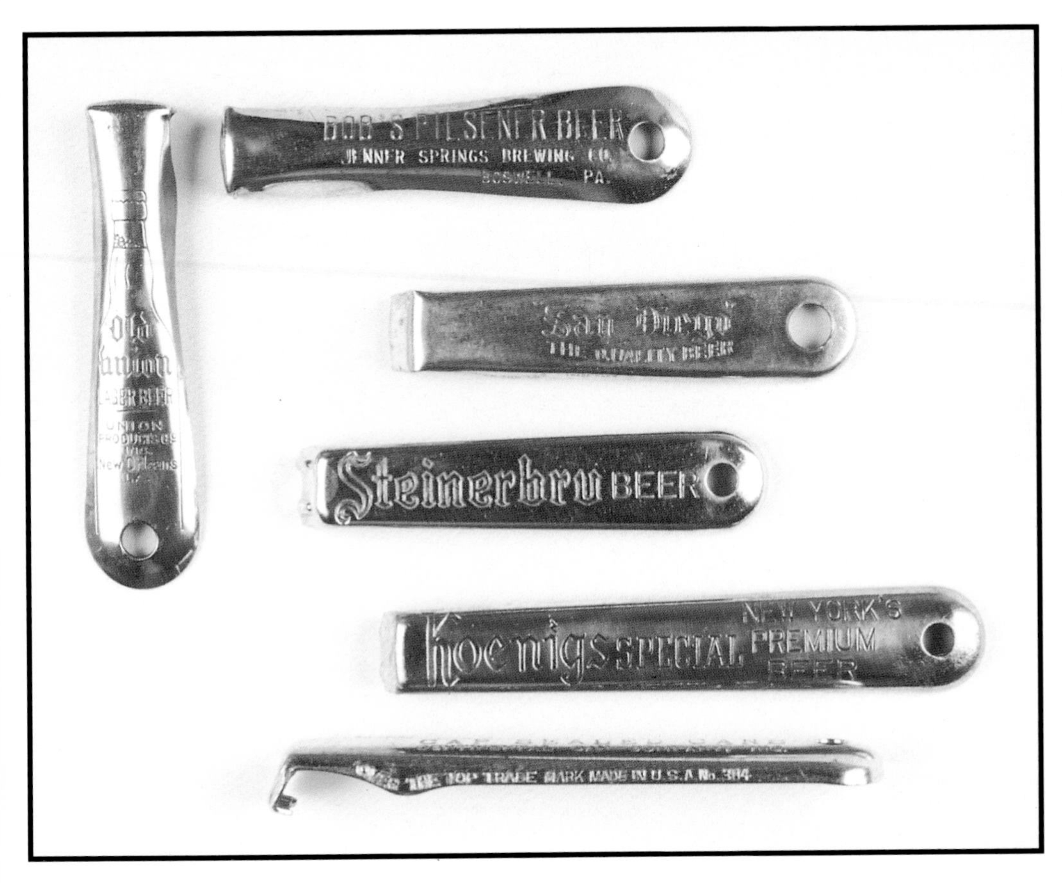

Left and top: H-1 Over-the-top cap lifte Widening contoured handle. American design patent 1,490,149 issued to Harry L. Vaughan, April 15, 1924. 3 1/2". $3-20. *National Beer of Baltimore simply stated on this type: "Tastes imported."* **Middle pair:** H-2 Over-the-top cap lifte It has a pair of curvilinear rounded point or sharp points, bearing edges that engage top surface of bottle cap. Marke VAUGHAN NOV. MFG CO. CHICAGO PAT.PEND. or VAUGHAN CHICAGO. American patent 1,490,149 issued to Harry L. Vaughan, April 15, 1924. 3 3/8 $3-20. *Playing off of the "It's the water" famous slogan from Olympia Brewing Company, Salem Beer simply countered with "It's the beer." Garden Ale used this reassuring advertising: "Good for what a you."* **Bottom pair:** H-3 Over-the-top cap lifter. It has a pair of curvilinear rounded points or sharp points, bearing edges that engage top surface of bottle cap. Marked VAUGHAN CHICAGO. American patent 1,490,149 issued to Harry L. Vaughan, April 15, 1924. 4 1/4 $3-20. *Continental Can Company promoted beer in general by using this opener with the statement "Beer is bette in cap sealed cans."*

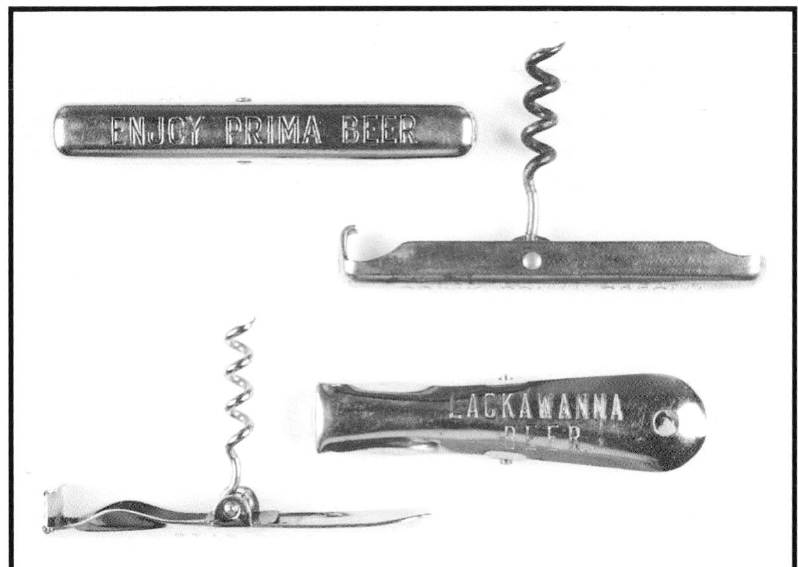

Top pair: H-4 Over-the-top cap lifter with folding corkscrew. Marked PATENTED DEC.11.28 FEB.12.29 (JEM CO IN LOGO) THE J. E. MERGOTT CO. NEWARK, N. J. American patents 1,695,098 and 1,701,950 issued to William Hiering, December 11, 1928, and February 12, 1929. 3 3/4". $10-30. **Bottom pair:** H-10 Over-the top cap lifter with folding corkscrew. Widening contoured handle. 3 1/2". $30-50.

Top to bottom: (pair) H-5 Combination over-the-top cap lifter and conventional opener. 4". $5-10. **(pair)** H-6 Over-the-top cap lifter. One large dimple. 4 1/4". $5-8. *On one example of this type Blatz Brewing Company says that Blatz is "Milwaukee's* ***Famous*** *Beer" and on another "Milwaukee's* ***Finest*** *Beer."* H-7 Over-the-top cap lifter. Two dimples. 4 1/8". $5-10. *On this one Blatz says their beer is "Milwaukee's Most Exquisite Beer."* H-9 Over-the-top cap lifter. 3 5/8". $60-75. **Right:** H-11 Over-the-top cap lifter. One dimple. 4 1/8". $5-10.

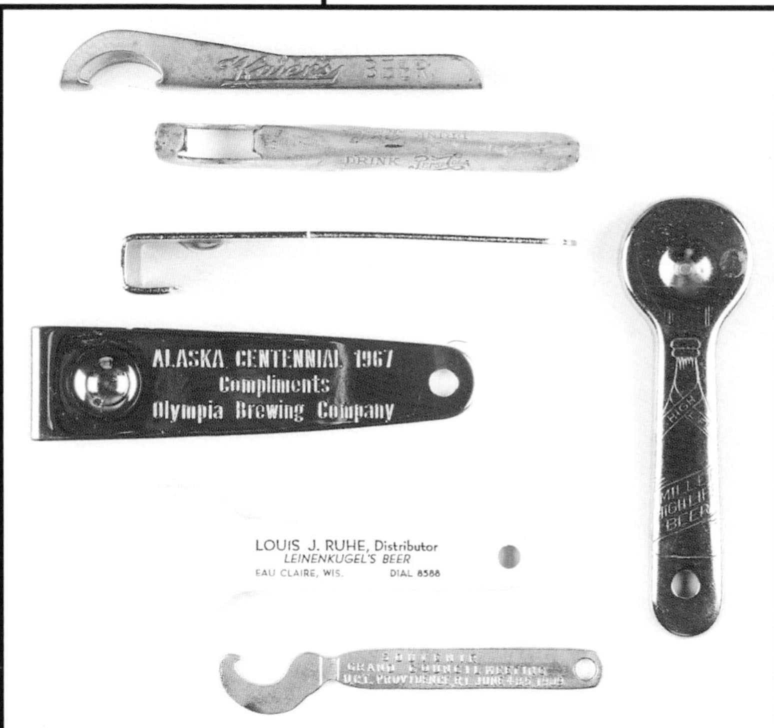

H-8 Over-the-top cap lifter. Bottle shape. Several advertising brands found. Cap lifter is at the top, except for "Pabst" which is at the bottom. All except "Pabst" and "Schlitz" marked on reverse: MUTH BUFFALO, N.Y. COPYRIGHT 1940 PAT PENDING. 4 1/8". $5-40.

Combination Cap Lifters/Can Piercers

On January 24, 1935, the first beer in cans was produced. The beer was from the Krueger Brewing Company, Newark, New Jersey. The can was manufactured by the American Can Company. The can piercer had already been invented in 1932 for opening other containers with liquid, but the demand now was significantly increased. Can piercers became a necessity of life and were given away by the millions by vendors of beer and other drinks in cans. Initially, American Can Company manufactured openers in Newark, New Jersey. From 1935 to 1936, they licensed Vaughan Manufacturing to produce openers as well. During the ensuing boom years for the can opener, Vaughan was by far the largest producer. Others included Handy Walden, Ekco, Emro, Mira, Crown, and Greene.

Basic types of cap lifter/can piercer combinations are shown in this category. Variations of the types include: slight differences in length and width; presence of "ears" designed to prevent dropping the opener into a bottle; bottle and can ends in opposite planes; presence or absence of a hanging hole; and presence or absence of strengthening ribs. Other variations include manufacturer's names, manufacturing years, patent dates, and patent numbers. Some of the openers are copper, nickel, cadmium, chromate, or brass plated.

Product names for this type include Quick and Easy, Can Tapper, Safe-Edge Can Piercer, Tu-way, Easi-Ope, and Por Ezy. An early American Can Company opener advises: "Don't throw me out! I'm the quickest and best opener for all liquid foods."

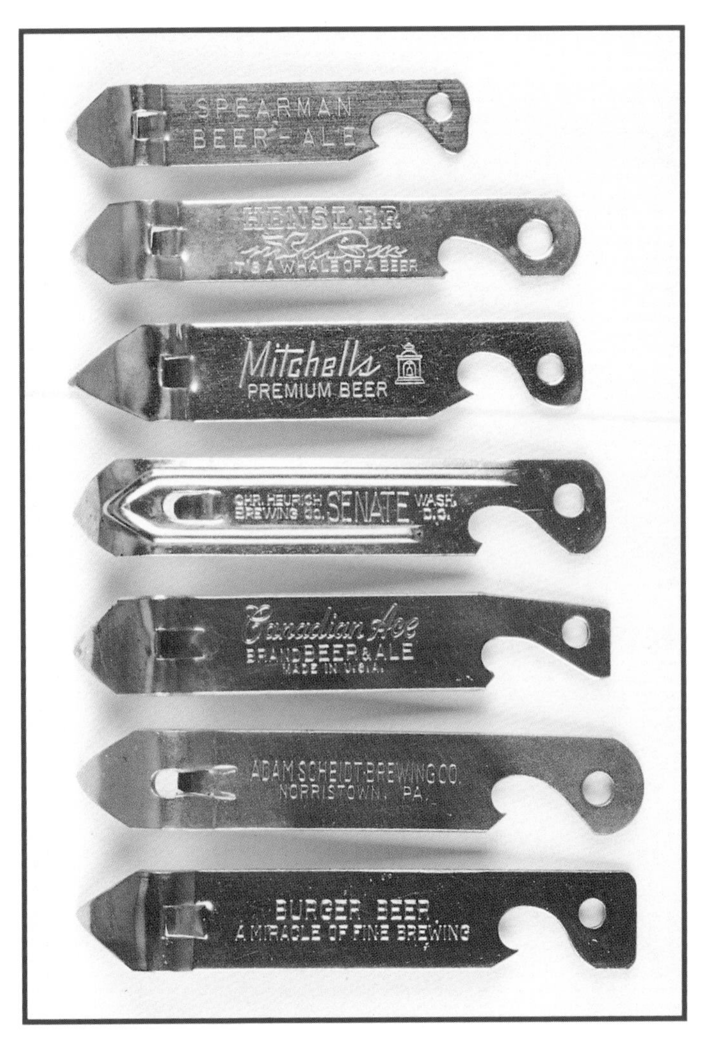

Top to bottom: I-1 Cap lifter/can piercer. 3 1/4". $1-5. *One example says "Emro Can Tapper, St. Louis, Mo."* I-2 Cap lifter/can piercer. 4" long and 5/8" wide. $1-5. I-3 Cap lifter/can piercer. 4" long and 3/4" wide. $1-5. I-4 Cap lifter/can piercer. American design patent 143,327 issued to Michael J. LaForte, December 25, 1945. 4 1/8". $1-5. *Bragging about their brands on this type are: "Arrow Beer, it hits the spot;" "For the best, buy Drewrys;" "Edelbrew grows its' own flavor;" "Harvard has what it takes;" "Schaefer, Beer at its best;" and "Sterling - The beer drinkers beer."* I-5 Cap lifter/can piercer. 4 1/4". $1-5. *On this type "Silver Top" claims to be "the most luxurious light beer in the world."* **(pair)** I-6 Cap lifter/can piercer. 4 1/2" to 4 5/8". Two variations shown. $1-5.

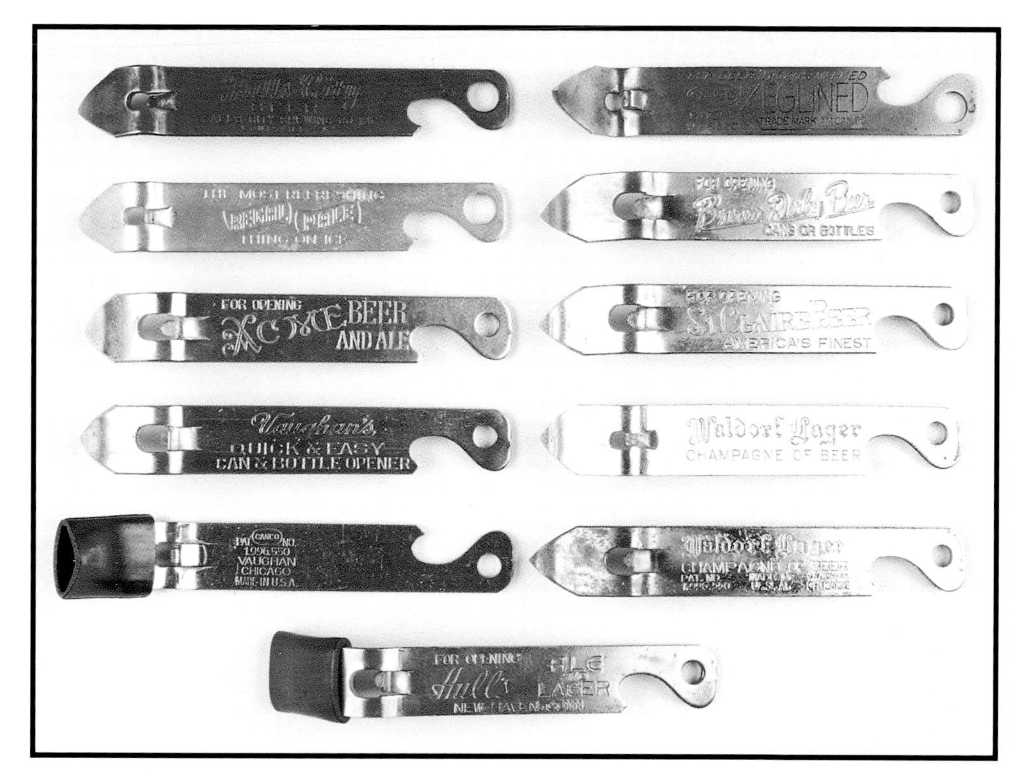

I-7 The original cap lifter/can piercer. Marked FOR BEER IN CANS MARKED KEGLINED CANCO PATENT 1,996,550 TRADE MARK AM. CAN CO. or PAT. NO. 1,996,550 VAUGHAN CHICAGO MADE IN USA or QUICK & EASY OPENER CANCO PATENT 1,996,550 MADE IN U. S. A. American patent 1,996,550 issued to Dewitt F. Sampson and John M. Hothersall, April 2, 1935. 4 3/4". $1-25. *The first canned beer was New Jersey's Krueger. They offered a choice on this type opener: "Enjoy Krueger's on draught. Where not available, drink it from Keglined cans."*

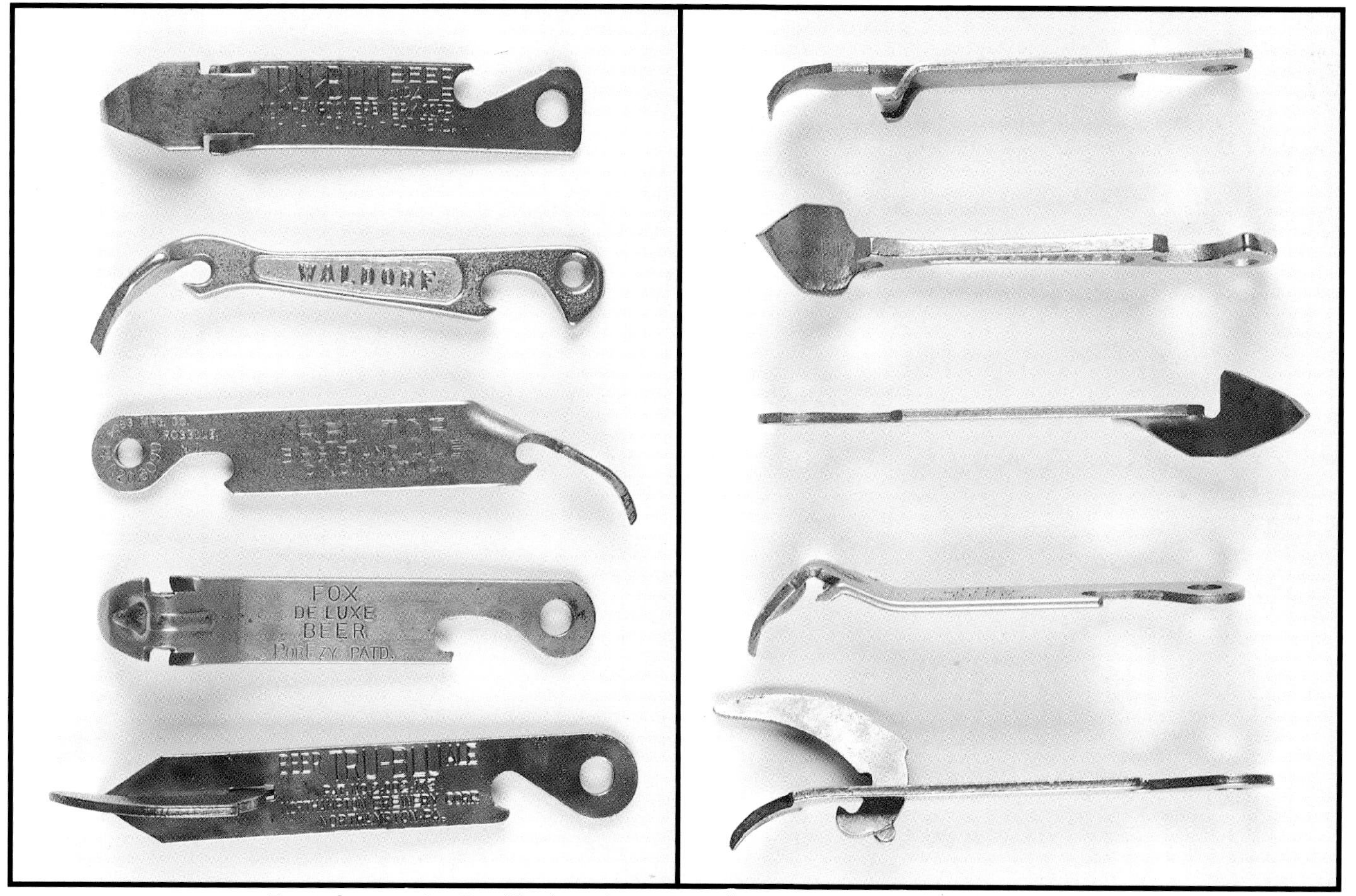

Top to bottom: I-8 Cap lifter/can piercer. Two ribs at can piercer end. 4 1/4". $15-20. I-9 Cap lifter/can piercer. Cast iron. 4 5/8". $10-30. I-10 Cap lifter/can piercer. The cap lifter is turned 90 degrees. Made by Soss Manufacturing Company, Roselle, New Jersey. American patent 2,019,099 issued to Francis H. Schwartz, October 29, 1935. 5". $20-25. I-24 Cap lifter/can piercer. Marked POREZY PATD. 4 3/8". $20-25. I-25 Cap lifter/can piercer. Two piece opener, movable member inserted in can piercer. Tooth of member grips flange of can. American patent 2,002,173 issued to Bernard E. Dougherty, May 21, 1935. 5". $75-100.

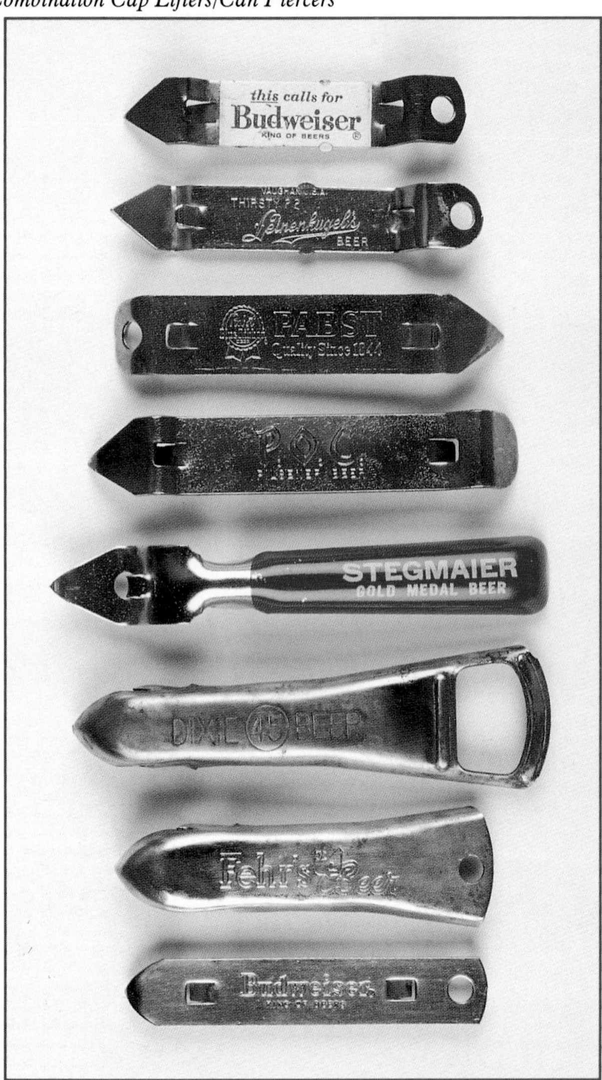

Left: **Top to bottom:** I-11 Cap lifter/can piercer. 3 1/4" to 3 3/8". $1-5. *Can you identify some of the symbols appearing on this opener? Eagle in letter A, three rings, tuba player, mountie, town crier, arrowhead, lion. (They are: Anheuser-Busch, Ballantine, Bosch, Drewrys, Knickerbocker, Oshkosh Brewing, and Gluek).* I-12 Cap lifter/can piercer. 3 5/8" to 3 3/4". $1-5. *Ready for a few more advertising symbols? Man on high-wheel bicycle, bear, rooster, and crown. (They are: Gretz, Hamm's, Goebel, and Kingsbury).* I-13 Cap lifter/can piercer. 3 7/8" to 4 1/8". $1-5. *Several of these were produced to mark milestones for the Peter Hand Brewery Company of Chicago: 250 million glasses yearly, 300 million glasses, and 450 million glasses.* I-14 Cap lifter/can piercer. 4 1/4" to 4 3/8". $1-5. *This opener was a souvenir of the 1957 Brewers Association of America convention (see also E-11).* I-21 "Tu-Way" Cap lifter/can piercer. Cap lifter and can piercer on the same end. 5". $2-10. I-22 "Easi-Ope" Cap lifter/can piercer. American design patent 164,448 (September 4, 1951) and mechanical patent 2,517,443 (August 1, 1950) issued to Harland R. Ransom. 4 3/4". $1-5. I-23 Can piercer. Same as I-22 but no cap lifter. 3 3/4". $2-5. I-28 Can lifter/can piercer. Flat - probably pulled off the production line before the ends were bent (see I-11). 3 5/8". $2-3.

Below left: **Top to bottom: (pair)** I-15 Cap lifter/can piercer. American patent 2,053,637 issued to Herbert Schrader, September 8, 1935. 4 3/4". Two different widths are shown. $1-5. *A simple statement appears on one for Old Reading Beer—"It's wonderful."* I-16 Cap lifter/can piercer. Ribbed. 3 7/8" to 4". $1-5. I-17 Cap lifter/can piercer. Ribbed. Some marked VAUGHAN PATENT 1,996,550. 4 1/4" to 4 3/8". $1-5. *A coupon campaign was promoted on this type with "Tavern Pale, Free gifts! Save those coupons" (with a picture of a hand removing a neck label from a bottle).* I-18 Cap lifter/can piercer. Ribbed. 4 3/4". $1-5. I-26 Cap lifter/can piercer. Ribbed. No hanger hole. Made by Campello, Massachusetts. 3 3/4". $8-10.

Below right: **Top to bottom:** I-19 Cap lifter/can piercer. Quad-fold. 3 1/2". A sales card from Vaughan, Chicago, describes this as "4 openers in 1. Can tapper, bottle opener, seal cutter, and vacuum lid opener." $1-5. I-20 Cap lifter/can piercer. Quad-fold. American design patent 168,053 issued to Michael J. LaForte, October 28, 1952. 3 7/8". $1-5. I-27 Cap lifter/can piercer. Quad-fold. Steel swivel ring (I-19 & I-20 have plastic). 3 3/4". $1-5. I-30 Cap lifter/can piercer. Plastic insert in handle. 4 5/8". $2-3.

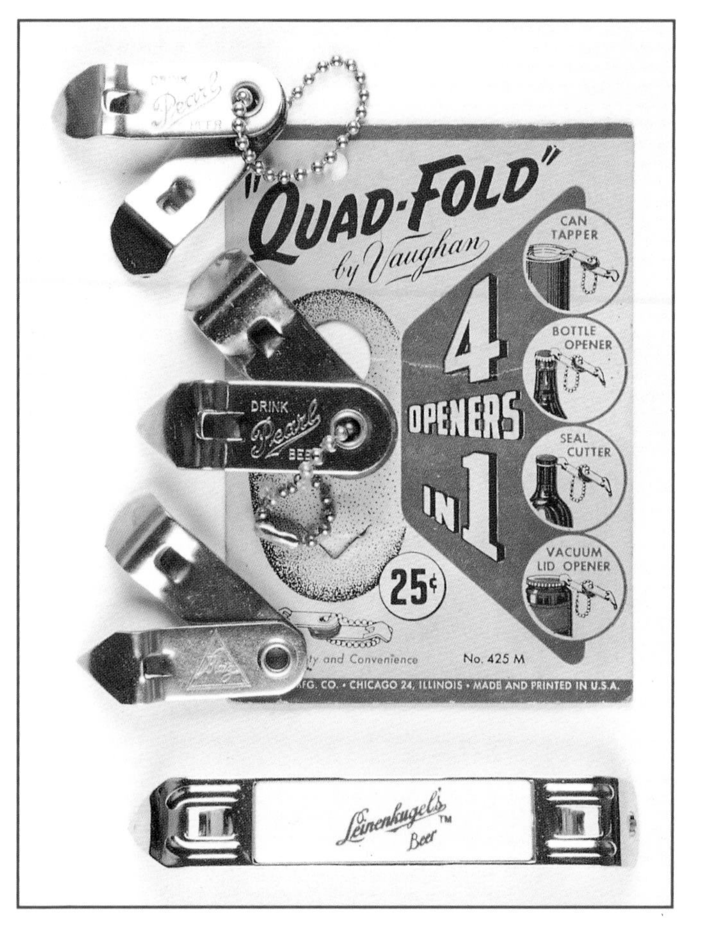

Can Piercers

Pat Stanley best described this type with these words: "If you know the 'I' type, you know what these look like. They are one-half of an 'I'."

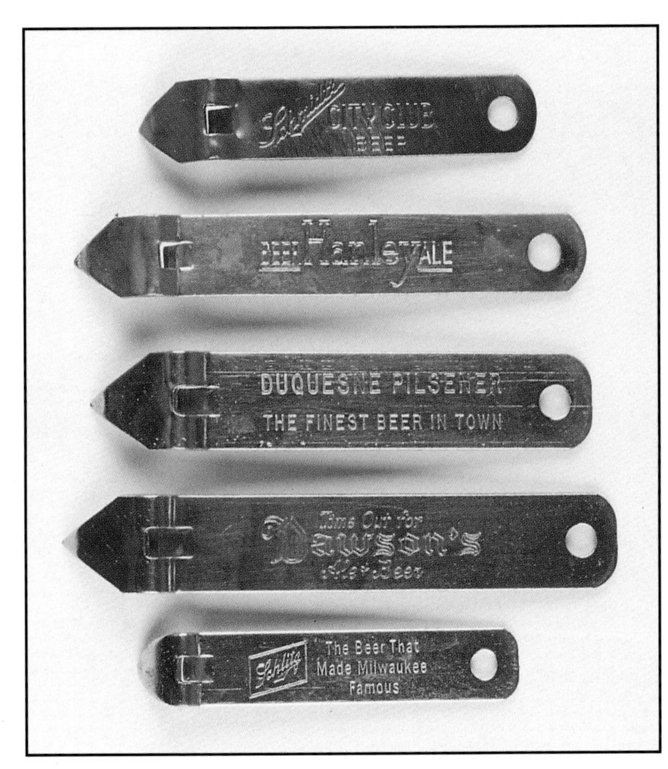

Top to bottom: J-1 Can piercer. Made by Walden, Cambridge, Massachusetts. 3 1/8" long and 5/8" wide. $1-5. J-2 Can piercer. 4 1/8" long and 5/8" wide. $1-5. J-3 Can piercer. 4 1/8" long and 3/4" wide. $1-5. J-4 Can piercer. 4 3/8" to 4 1/2". $1-5. J-11 Can piercer. Vaughan Company's "O-G Junior Can Tapper." 2 7/8" long and 9/16" wide. $1-5.

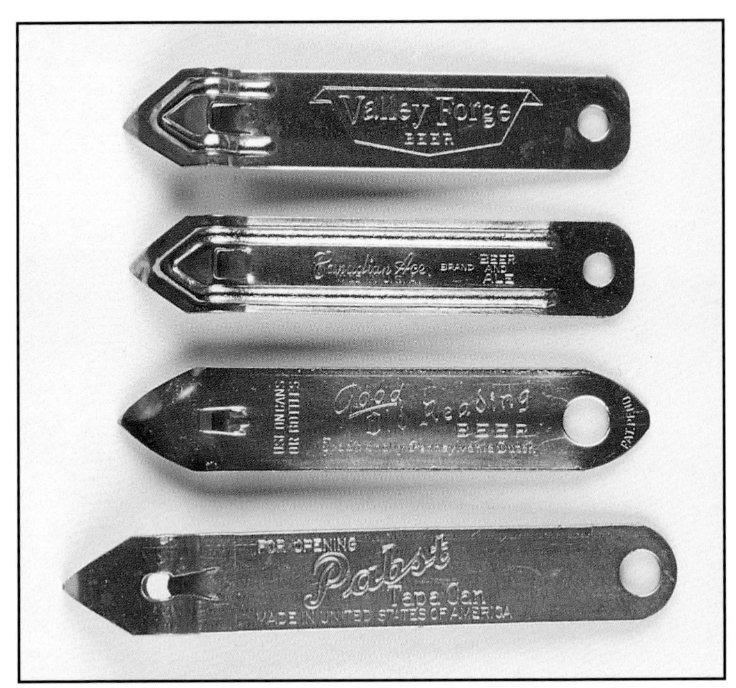

Top to bottom: J-5 Can piercer. Ribbed can piercer end. Made by Vaughan, Chicago. 4". $1-5. *Tivoli Beer expressed regionality on this type with "Brewed for Western tastes."* J-6 Can piercer. Whole opener ribbed. Marked PAT. 143,327; 1,996,550, OTHERS PENDING, VAUGHAN, CHICAGO, MADE IN U. S. A. 4". $1-5. J-8 Can piercer/cap lifter. Marked USE ON BOTTLES OR CANS or HARRAH, U. S. A., PAT. 1,996,550 OTHERS PEND. or PAT. PEND. The patent marking is Sampson and Hothersall's 1935 patent, yet this is American patent 2,773,272 issued to George R. Harrah, December 11, 1956. 4 1/4". $1-5. J-10 Can piercer. Like I-7 but no cap lifter. Marked FOR BEER IN CANS MARKED KEGLINED CANCO PATENT 1,996,550 TRADE MARK AM. CAN CO. 4 3/4". $10-12. *The only advertising found for this one is "For opening Pabst Tap-A-Can, Made in United States of America, for beer in cans marked Keglined."*

Top: J-9 Can piercer with tubular handle. American design patent 155,314 issued to Joseph G. Pessina, September 20, 1949. 4". $1-5. **Bottom left:** J-7 Folding can piercer. Marked FOR BEER CANCO, PATENT 1,996,550. American patent 2,188,352 issued to Dewitt F. Sampson and John M. Hothersall, January 30, 1940. The opener is stamped with Sampson and Hothersall's 1935 patent number. 4 3/8". $5-20. **Bottom middle:** J-12 Can piercer. Picture of bottle. 2 5/8". $8-10. *This one gets the award for the most patriotic advertising with "Busch Bavarian Beer, Eternal vigilance is the price of liberty. If we understand the American way, we won't accept socialism, communism, or unlimited government (American Flag)."* **Bottom right:** J-13 Folding can piercer. 3". $30-40.

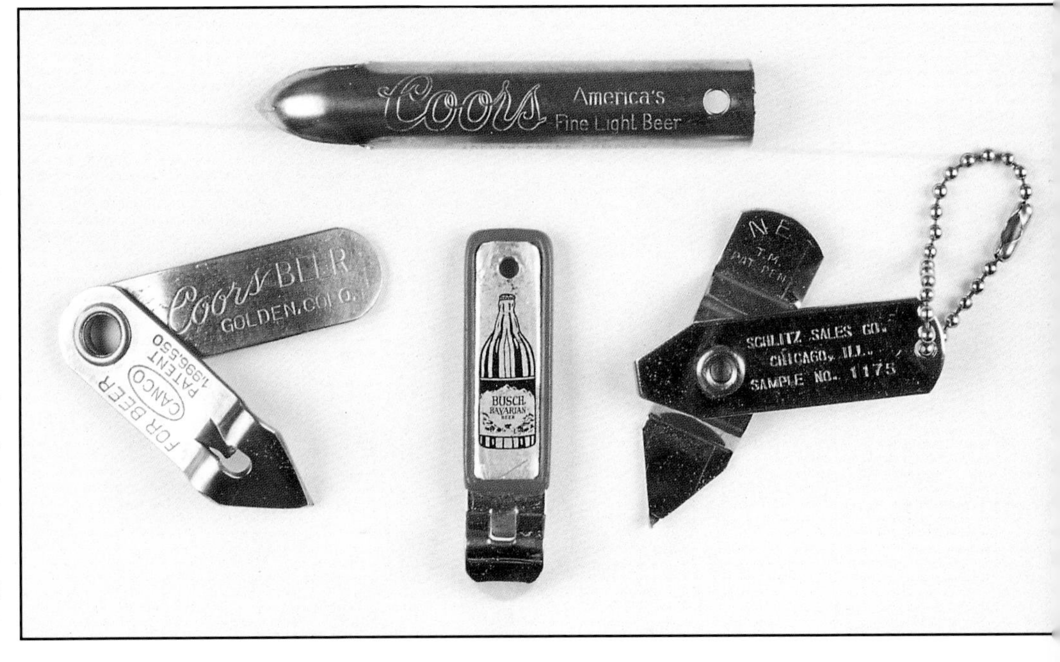

Special Can Piercers

Before capped beer bottles and beer cans, beer bottles were corked. One of the best inventions for the hurried bartender was the bar mounted corkscrew (see corkscrew section). These corkscrews enabled the bartender to rapidly and efficiently remove corks from beer bottles in a single stroke. The bartender, faced with cases of cans to open for thirsty customers, can quickly tire of using the hand held can piercer. The counter top can piercer was the perfect solution. The bartender could simply slip the can up to the back of the opener face-plate, lower the handle to pierce the can, and serve the thirsty customer. The Retz Company of Fort Worth, Texas, described it best by advertising their can opener as the "One Shot Can Opener."

Left to right: (pair) K-2 "One Shot Can Opener." Pot metal opener for 12 oz. Cans with double can piercer. Holes in base for mounting to bar. A sales flyer from the Retz Company, Fort Worth, Texas, indicates they are the sole distributors. 5 7/8". $30-50. K-14 same as K-2 but for 16 oz. cans. 7 1/2". $50-60.

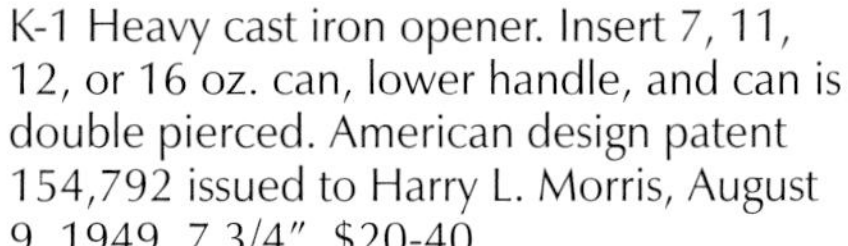

K-1 Heavy cast iron opener. Insert 7, 11, 12, or 16 oz. can, lower handle, and can is double pierced. American design patent 154,792 issued to Harry L. Morris, August 9, 1949. 7 3/4". $20-40.

K-3 "Can Handle." Combination can piercer and handle (for 12 oz. can). Manufactured in 1960 for the Rheingold Brewing Company by the Handy Walden company. 5". $10-15. K-4 "Tapster." Open lid, insert can, close lid that pierces can, then pour. Made by Revere Copper and Brass of Rome, New York, 1936-1941. 5 3/8". $200-300.

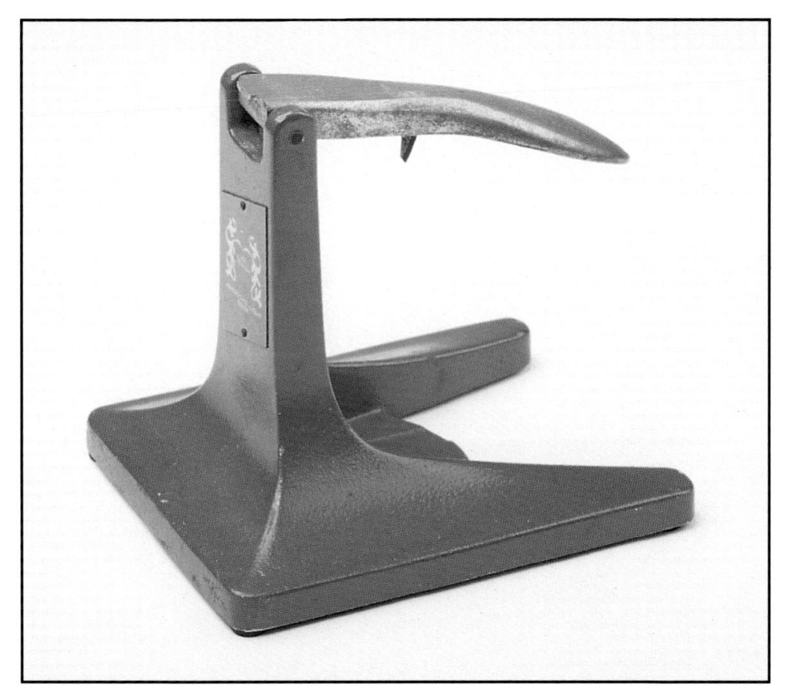

K-5 Double can piercer for 12 oz. can only. Cast iron version by Vaughan, Chicago. Cast aluminum version made by Andy Wadoz & Co., Milwaukee. 6". $20-60.

K-6 Unusual counter top can piercer. The can rests on a platform that swings up to punch the can. 7 3/4". $100-125.

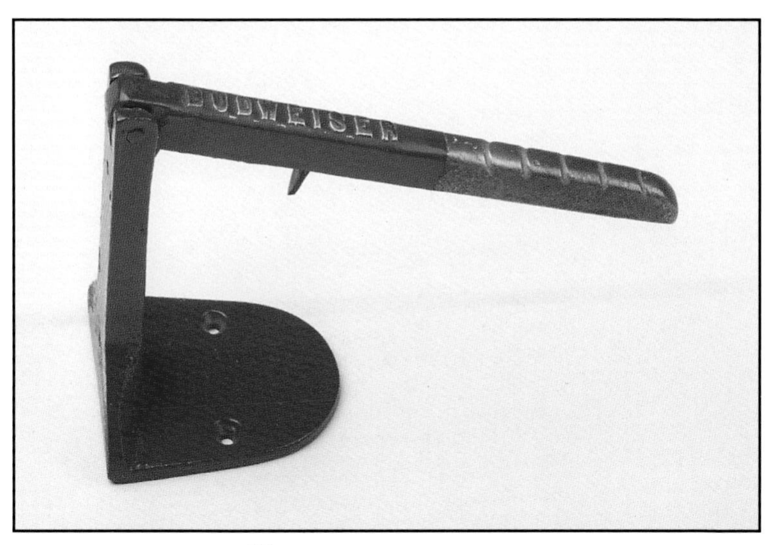

K-7 Cast aluminum can opener with double can piercer. Marked BAY FOUNDRY COMPANY TAMPA. Has holes for mounting on bar or on wall. 5 1/2". $150-200.

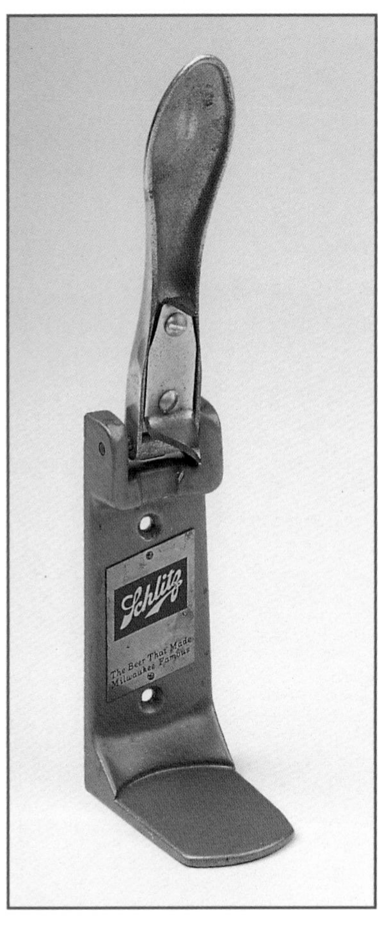

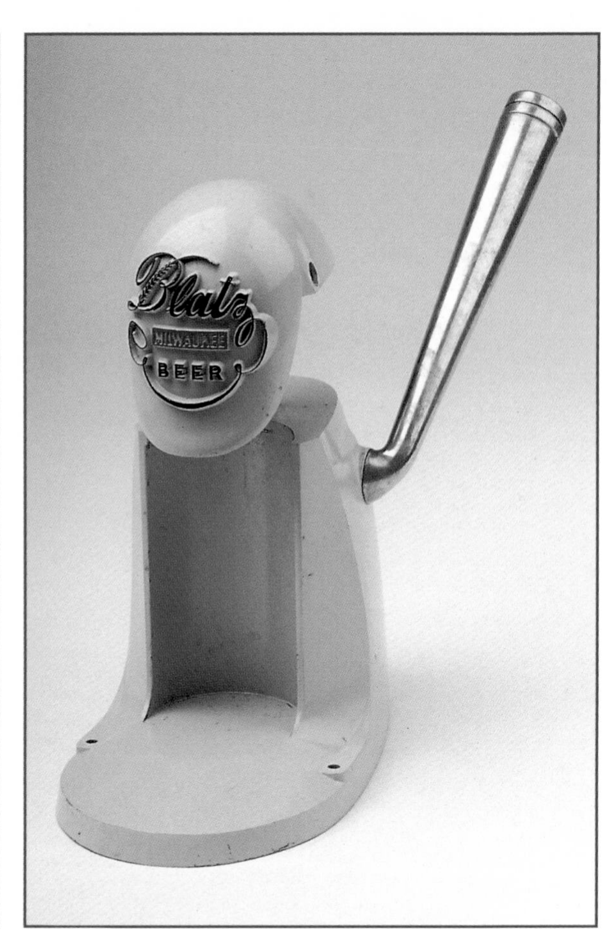

Right: K-8 Wall mounted double can piercer. 5 3/4". $40-50.

Far right: K-9 Can opener. Cuts into top so can serves as a drinking glass. 7 3/4". $100-150.

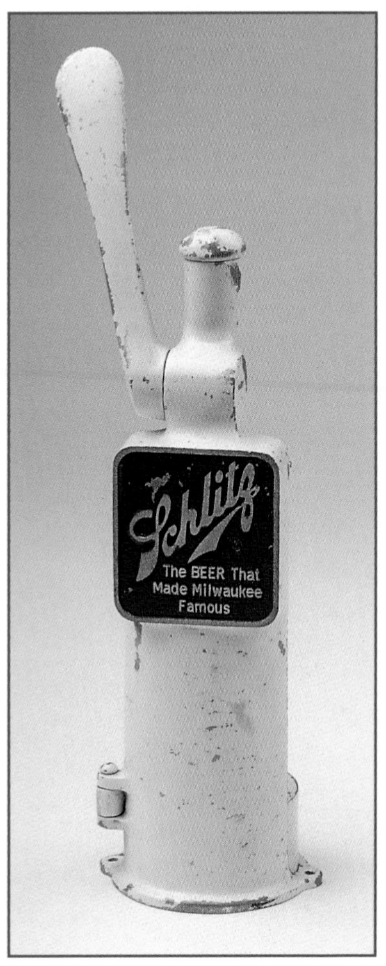

Two at right: K-10 Cast aluminum bar mount can piercer. Made by Ross Aluminum Foundries Company of Sidney, Ohio. On one version a hockey puck like piece is inserted for opening short cans. On another, there is a pivoting shelf for opening tall and short cans. 9 3/4". $150-200.

Above: K-11 Bar mount can piercer. Marked PAT. 2569123. 13 3/4". $75-100.

Above right: K-12 Can opener with cut out of Texas shape. American design patent 181,856 issued to Eric Johnson, January 7, 1958. 13 1/2". $50-75.

Right: K-13 Heavy cast iron can piercer. Double pierces 16 oz. cans only. 8 1/2". $50-75.

Can & Bottle Shape Openers

The earliest bottle shape openers were corkscrews. The worm was stored inside of a two part bottle which served as the pulling handle to remove the cork. About two dozen brewers had tablets attached to these bottles to promote their products. But what better way for a brewer to promote his beer, than by replicating the whole can or bottle in miniature form? In the 1930s and 1940s, the mini-bottle fad was in full swing in the form of salt and pepper shakers, liquid filled glass bottles, and wood bottles with reproductions of labels. The wood bottles often had a cap lifter attached. Blatz, Budweiser, Coors, Miller, Pabst, Ruppert Knickerbocker, and Schlitz later followed with miniature versions of their canned beers. These free standing three-dimensional bottles and cans add a lot of color and character to any collection of openers.

L-1 Retractable can piercer. The piercer is released by pushing a button on the top of the can. Made in West Germany. 1 3/4". $10-30.

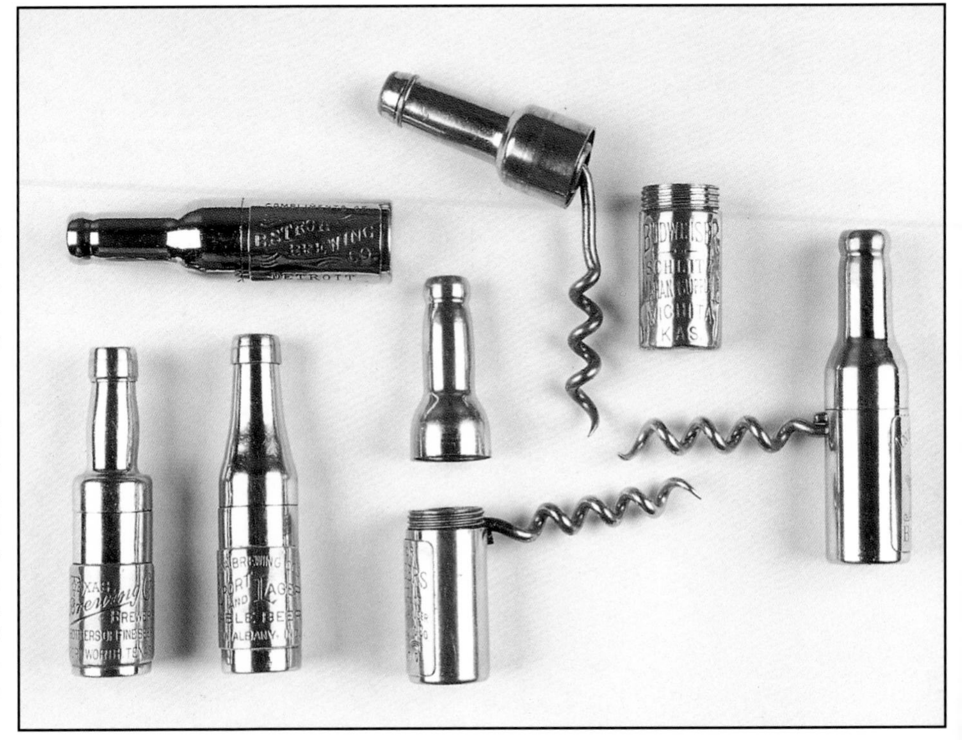

L-2 Mini bottle roundlet corkscrew. Round or square shoulders on bottle. American patent 583,561 issued to William A. Williamson, June 1, 1897 for his invention of a corkscrew concealed inside a small bottle or bullet shape roundlet. The ends thread together and, when unscrewed, a helix pivots at an angle to the base. The two pieces are then screwed back together to form the handle. Advertising plaques were applied to some bottles and bullets. American patent 657,421 issued to Ralph W. Jorres, September 4, 1900, for his version of this corkscrew. Jorres attached the helix to the top of the bottle instead of the base. Bottles were produced by the Williamson Company in nickel plated brass. Some bottles have Stanhopes in the top. The top is held up to the light and a small magnifier (Stanhope) contains a photograph. Unfortunately, the photo was delicate and finding one complete is rare. 2 3/4". $25-200.

Left to right: L-3 Michelob bottle made of wood and metal. An Anheuser-Busch promotional opener of the 1960s that sold for 84¢ at that time. 6". $15-20. L-5 Wood mini bottle with cap lifter attached. An example advertising "Chief Oshkosh Beer" has a corkscrew on one end of the cap lifter. The wood bottle is the sheath. 3 3/4". $30-40. L-8 Cap lifter with plastic bottle. 6". $15-20. L-9 Bottle with cap lifter. 6 5/8". $5-7

Left to right: L-7 Wood mini bottle with corkscrew inside. 6 1/2". $60-75. L-10 Wood bottle with over-the-top cap lifter. 8 1/4". $40-50. L-11 Wood bottle with wire cap lifter. Three frets. 6 1/8". $5-7. L-12 Paper label encased in lucite bottle and cap lifter. 6 5/8". $5-7. L-13 Steel mini bottle with combination cap lifter, corkscrew, and loop seal remover. 5". $125-150.

Left: L-14 Plastic bottle with cap lifter. 2 5/8". $3-5.
Middle: L-15 Half bottle with cap lifter. Marked PEWTER. 5". $10-12.
Right: L-16 Plastic bottle with over-the-top cap lifter. Marked MOD-DEP. 5". $3-5.

L-4 Over-the-top type cap lifter on mini-bottle. The wooden bottles were sold in the late 30s and early 40s for advertising giveaways and souvenir shop sales. Various heights and diameters. 3 1/2" to 4 1/4". $10-40.

Miscellaneous Openers

When the classification system for openers was presented in the 1978 book *Beer Advertising Openers*, there were 28 openers that didn't seem to fit in any of the other categories. There were bell shapes, cast iron openers, openers with stag handles, star shape, bullet shape, and even a couple with corkscrews. The "Miscellaneous" category became sort of a "dumping ground" for those perplexing pieces that raised the question "where the heck do I put these?" Now there are over 100 openers here and, although one could argue for classifications for luggage tag type, tableware, bone handles, and many others, we've opted to just leave the lot in this dumping ground. We invite readers to call them whatever they like!

M-1 Lithographed cap lifter. Rounded base. Made by H. D. Beach Company of Coshocton, Ohio. Some marked with patent date, Sept. 11, 1911. American design patent 41,807 issued to Harry L. Beach, September 11, 1911. Front side curves in. Back side curves out. 3 1/2". $25-150.

M-2 Lithographed cap lifter. Squared base. Made by H. D. Beach Company of Coshocton, Ohio. Some marked with patent date, Sept. 11, 1911. Front side curves in. Back side curves out. 3 1/4". $25-150.

M-3 "Prestopener." Retractable or slide-out opener made by Electro-Chemical Engraving Co., Inc. of New York. Produced in over-the-top type and conventional cap lifter type. The opener is steel and the casing is plated and painted over brass. Colors include black, blue, green, red, yellow, and orange. 2 1/2" closed. 3 1/4" to 3 1/2" open. $20-150. *Advertising on this type with a racing theme was used by Hialeah Brewing Company, Hialeah, Florida: "Jockey Club Beer, A sure winner."*

Type M-3 slide-outs and value ranges vary by brewery or brand name. Values are usually for full paint and, at most, a couple of minor inside corrosion spots. Color variations exist and may affect value slightly (for example: a green Edelweiss is more common than other Edelweiss colors). Mint condition can certainly raise the value. Values are:

• $100-150: Hialeah, Piel Bros, Queen City, Ruppert, Seitz, Sheridan, Trommers.
• $75-100: Deer Park, Erie, Flocks, Fuhrmann & Schmidt, Harrison Golden Brew, Harrison Heidelberg, Heurich, Moose, Standard.
• $50-75: American, Brackenridge, Duquesne, Gunther Shield.
• $40-50: Bismarck, Chartiers Valley, Cumberland, Lang, Schorr-Kolkschneider, Stegmaier, Victor Knight, Victor Trademark, Zetts.
• $20-30: Barbeys, Beverwyck, Croft, Dauefer-Liberman, Edelweiss, German, Gunther 1881, Philadelphia, Piels.

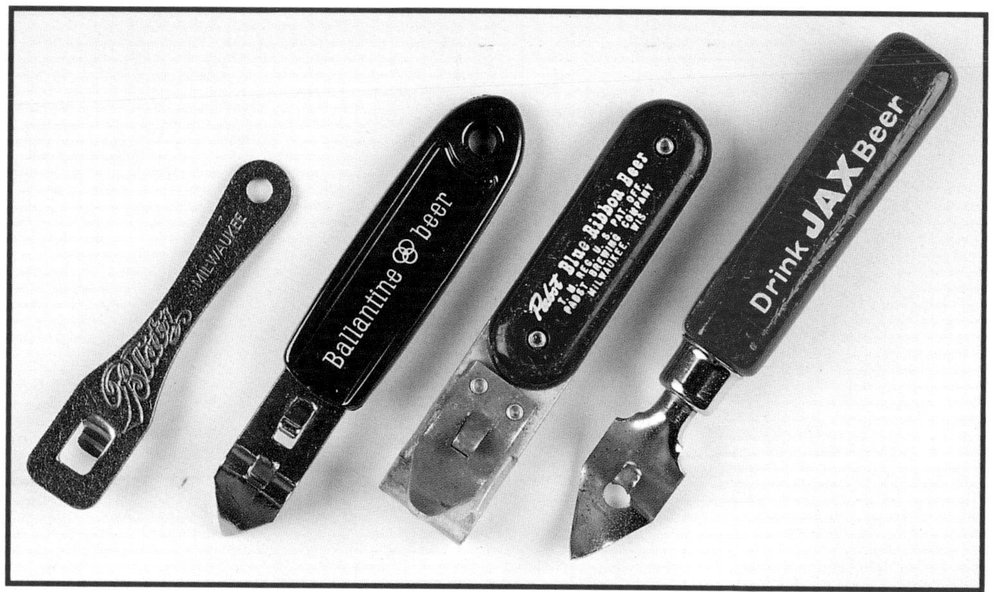

Left to right: M-4 Flat steel opener with cap lifter punched and pressed downward in two 90° bends to grip cap. Found only with "Blatz Milwaukee" in script and in block letters. 3 7/8". $3-5. M-9 Cap lifter and can piercer with plastic handle. Made by Vaughan Company. 5". $5-10. M-32 Over-the-top cap lifter and can piercer. Marked NU PRO CORP., PAT. APP. 211,085. Made by Nu Pro Corp. of Chicago, Illinois. 4 3/4". $5-8. M-33 Cap lifter and can piercer with wood handle. Marked EDLUND CO. INC. BURLINGTON, VT. U.S.A. 6". $15-20.

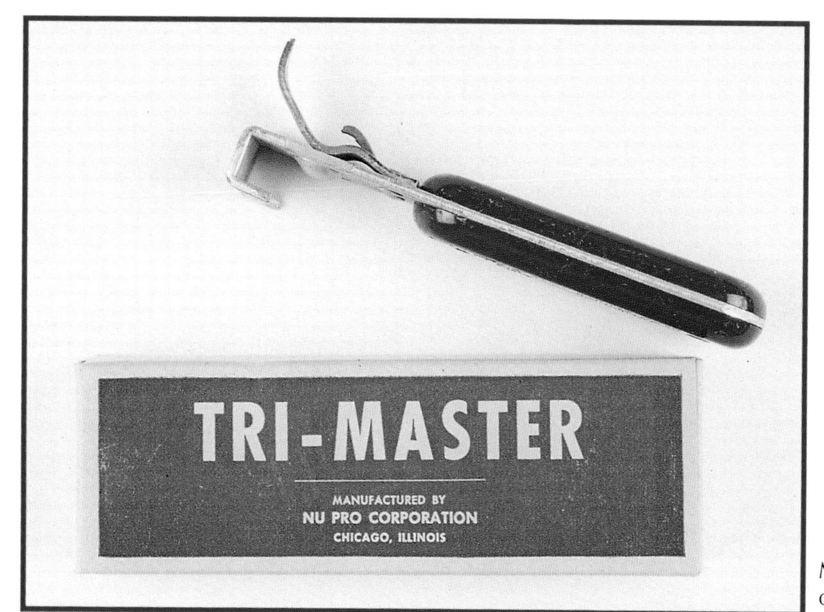

M-32 "Tri-Master" in original package.

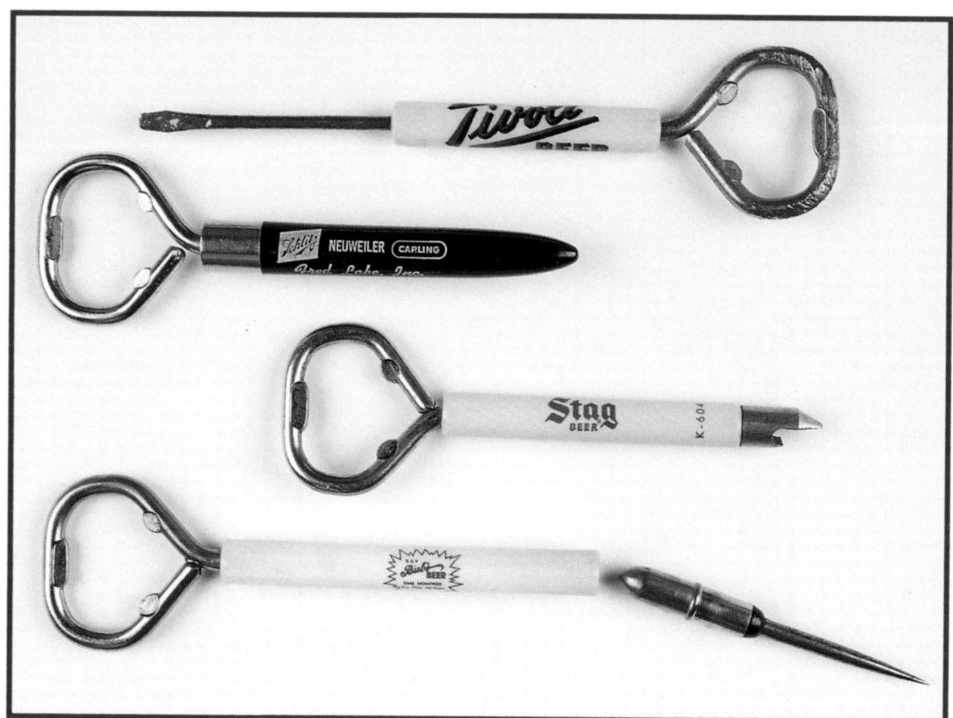

Top to bottom: M-5 Cap lifter with screwdriver. Celluloid handle. 6 1/2". $10-12. M-51 Cap lifter with plastic handle. Made by Newton Mfg. Co. of Newton, Iowa. 5". $8-10. M-64 Cap lifter and can piercer. Plastic handle. 5 1/8". $15-20. M-65 Cap lifter with ice pick. Ice pick point stores in plastic handle. It is pulled out, reversed, and inserted into plastic handle for use. 6 1/4". $15-20.

Left pair: M-6 Cap lifter with bell. Marked PAT. PENDING. 3 1/8". $30-75. *Hopfheiser, Krueger, Wagner, and Walter's used the bell to send their message "Ring for (beer name)." Signal Beer advertising told the owner to "Signal for a taste sensation."*
Right: M-37 Cap lifter with brass bell. Marked BELLS OF SARNA REGD U. S. 230 INDIA. 4 5/8". A 1949 Schlitz giveaway. $40-50.

M-37 Cap lifter bell in original box. The bell tag advertises: "This clever, practical bottle opener is handmade in far-off India. Every time you uncap a bottle, one bell will merrily tinkle announcing refill time. The life of the party in more ways than one, and a decorative 'note' to the bar. . . S. S. Sarna, New York 10, N. Y."

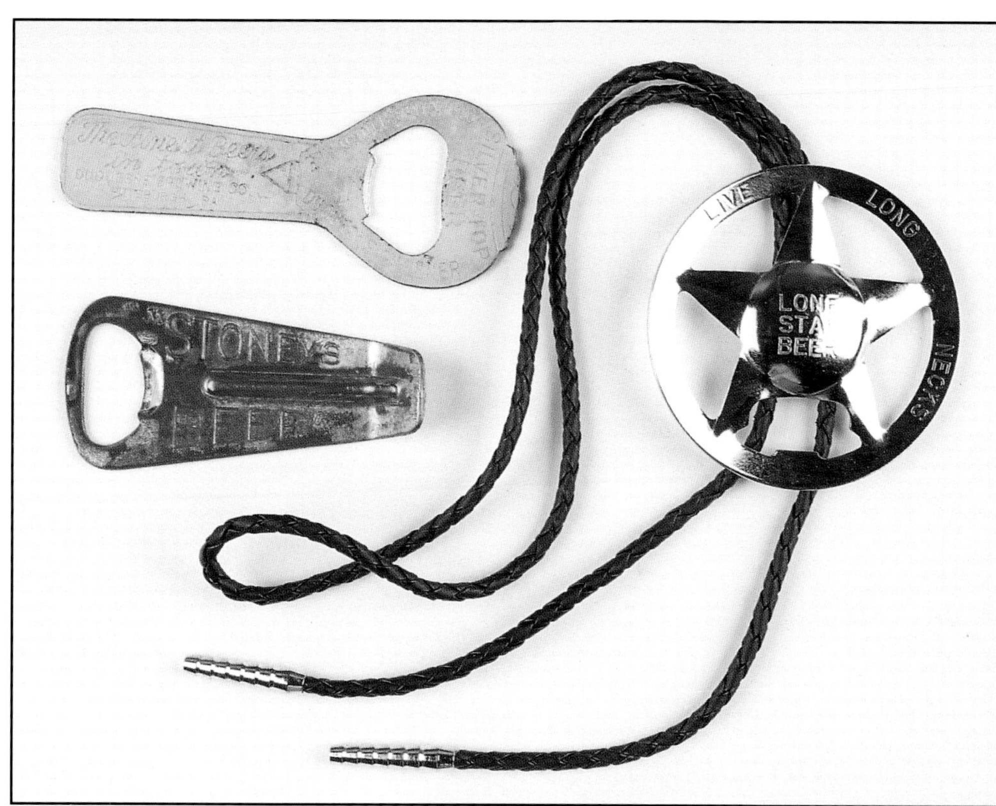

Top left: M-7 Flat steel cap lifter depicting a spinning top. 4 1/8". $20-25. *Made for Duquesne Brewing Co. of Pittsburgh advertising Silver Top Beer, Old Nut Brown Ale, Duquesne Pilsener.*
Bottom left: M-8 Cap lifter with handle bent at an angle. Has reinforcing rib. 3 1/4". $30-35. *Only found with advertising for "Stoney's Beer."*
Right: M-27 Cap lifter, key chain, and belt buckle or bolo tie. Featured in a 1975 brochure from the David Bortner Company, Janesville, Wisconsin. Works as a spinner for deciding who pays. Top point is labeled "You Pay" on the reverse side. 2 7/8". $15-25.

Advertising flyer for the Bortner Company's star spinner.

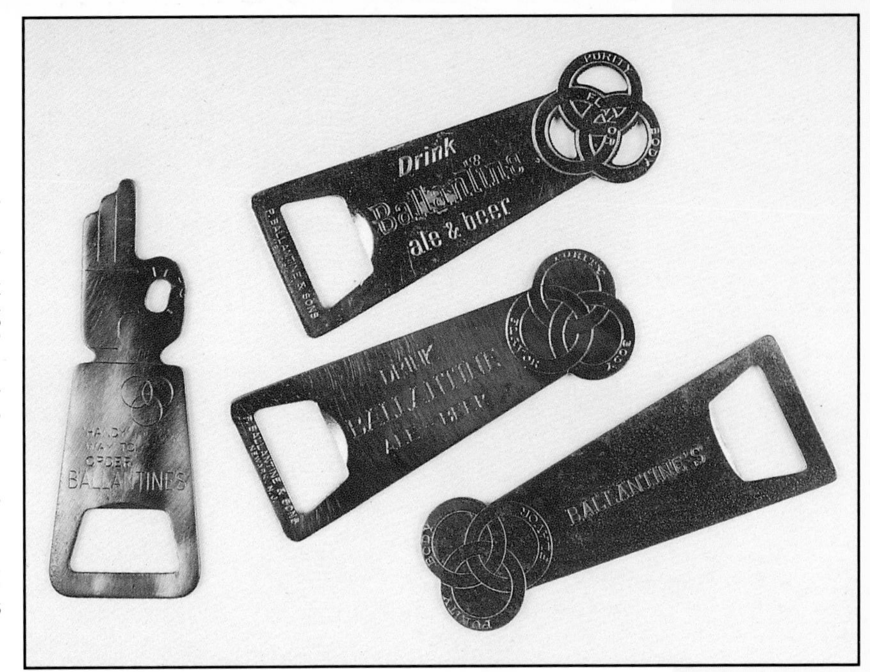

Left: M-10 Flat steel hand shaped cap lifter. Made by Vaughan Company. 4". $5-10. *Advertising opener for P. Ballantine & Sons, Newark, New Jersey: "'Handy' way to order Ballantine's."* **Right side, top to bottom:** M-11 Flat steel cap lifter with open three ring trademark (Purity, Body, Flavor) of P. Ballantine & Sons, Newark, New Jersey. 4 1/4". $2-4. M-12 Flat steel cap lifter with solid three ring trademark of P. Ballantine & Sons, Newark, New Jersey. 4". $2-4. M-13 Flat steel cap lifter with solid three ring trademark of P. Ballantine & Sons, Newark, New Jersey. Produced in right-hand and left-hand versions. When a right-hander holds a left-handed opener in his hand, the advertising is upside down (and vice-versa). 4". $2-4.

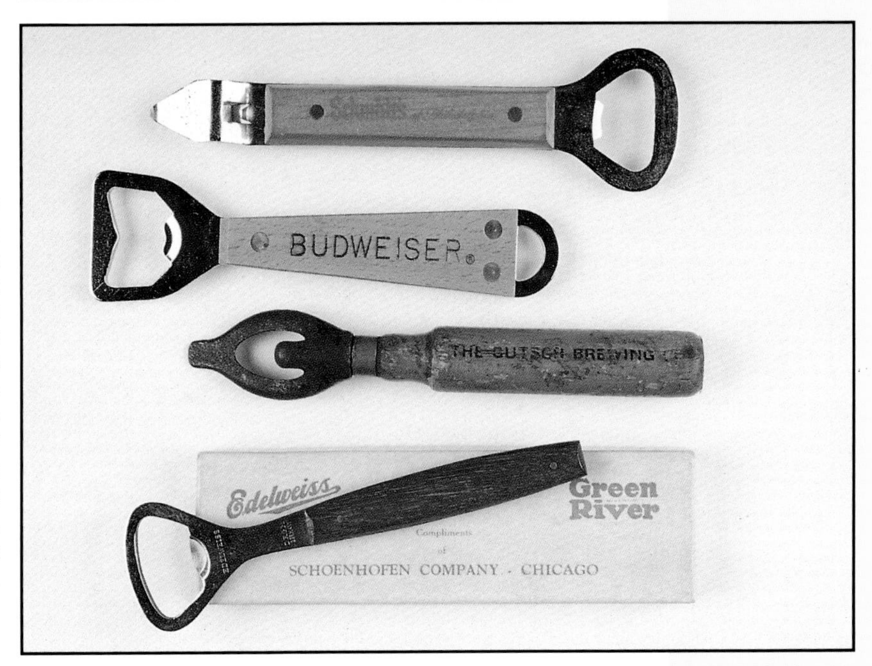

Top to bottom: M-14 Cap lifter and can piercer with wood handle. Made in Germany in 1963 as a gift for persons on tour through the Schmidt's Philadelphia plant. 6 1/8". $15-20. M-16 Cap lifter with wood handle. The handles of the openers were made from old Budweiser aging tanks and the openers were given to Anheuser-Busch executives at a special company meeting. They advertise "Budweiser/ Beechwood." 5 3/8". $15-20. M-20 Cap lifter with wood handle. 6". $40-50. M-22 Cap lifter with wood handle. Marked D. R. G. M., GERMANY (German design registration). 5 3/8". $40-50.

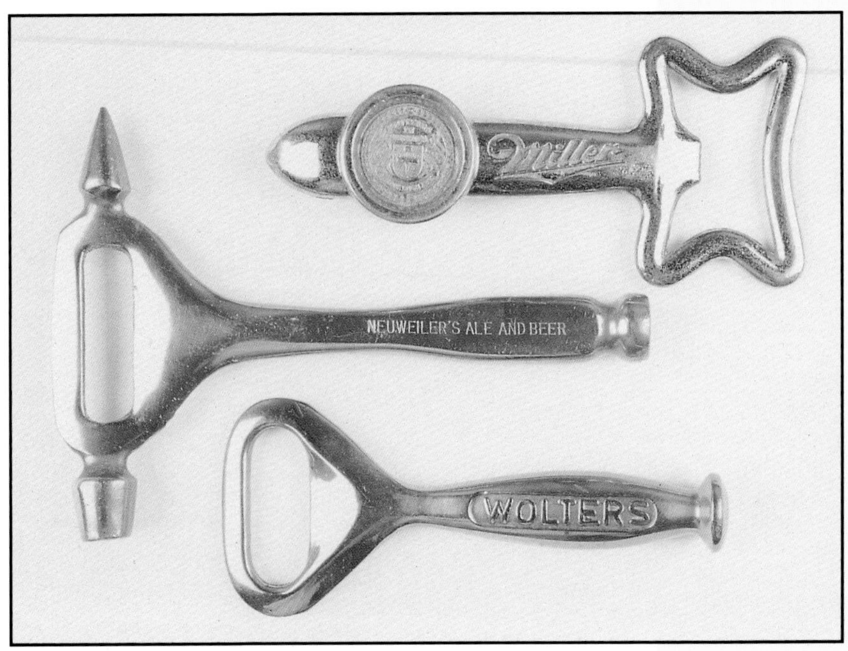

Top to bottom: M-15 Cast iron cap lifter and can piercer. Made by Federal Die Casting Co., Chicago for the Miller Brewing Company centennial celebration in 1955. 2000 of this type were made with the Miller name and brass plated. 2000 were made with the Federal name and chrome plated. 5 3/8". $20-25. M-84 Cap lifter, ice hammer, and ice pick. Marked PAT PEND U.S.A. 6 1/4". $60-75. M-19 Cap lifter and muddler. American design patent 148,535 issued to Frank E. Hamilton, February 3, 1948. 5 1/4". $10-35. M-58 Cap lifter similar to M-19 but with can opener at muddler end. 6 3/4". Not shown. $50-60.

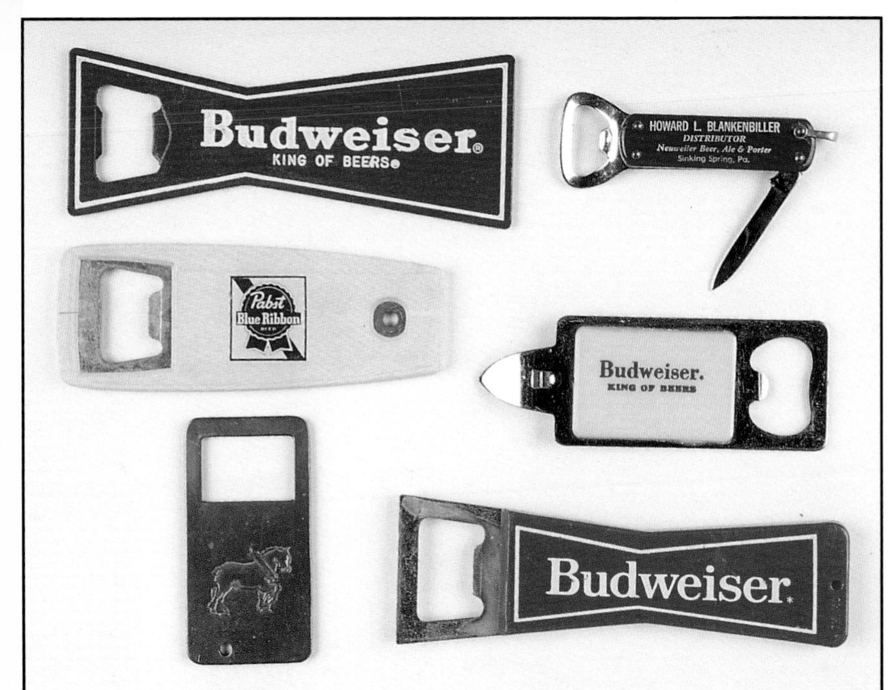

Left side, top to bottom: M-17 Opener in the shape of a bow-tie, an Anheuser-Busch Budweiser symbol. Advertising "Budweiser King of Beers" in white letters on red. Made in Japan. 5". $10-15. M-70 Cap lifter. Plastic handle. 4 1/2". $3-5. M-71 Cap lifter. Depicts Anheuser-Busch Clydesdale horse. Marked LOWELL SIGMUND INC. COPYRIGHT 1979. 2 7/8". $3-5.
Right side, top to bottom: M-72 Cap lifter with single blade knife. 3". $20-25. M-101 Cap lifter and can piercer. Marked PROV.CUT.CO. PROV.R.I.U.S.A. 4". $5-8. M-102 Cap lifter in the shape of an Anheuser-Busch Budweiser bow-tie. Marked TM/ MC. 5 1/2". $5-8.

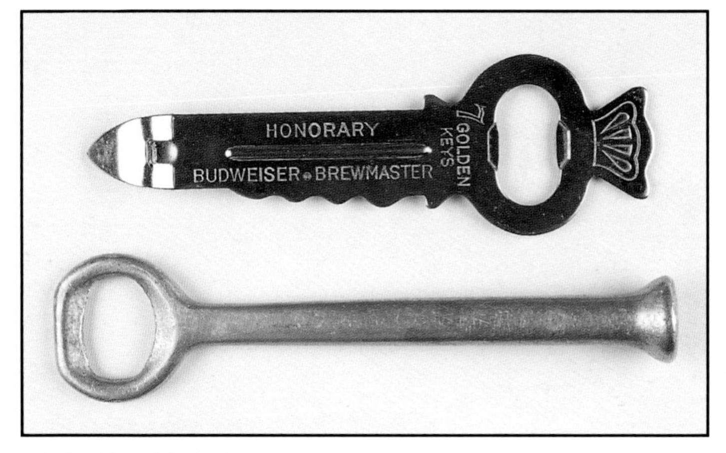

Top: M-18 Cap lifter and can piercer labeled "Honorary Budweiser Brewmaster, 7 Golden Keys." The Golden key was an advertising - promotion item during the 1950s. The seven keys refer to aspects of Budweiser. 6 1/2". $15-20.
Bottom: M-67 Cap lifter. Cast aluminum. 7". $60-75.

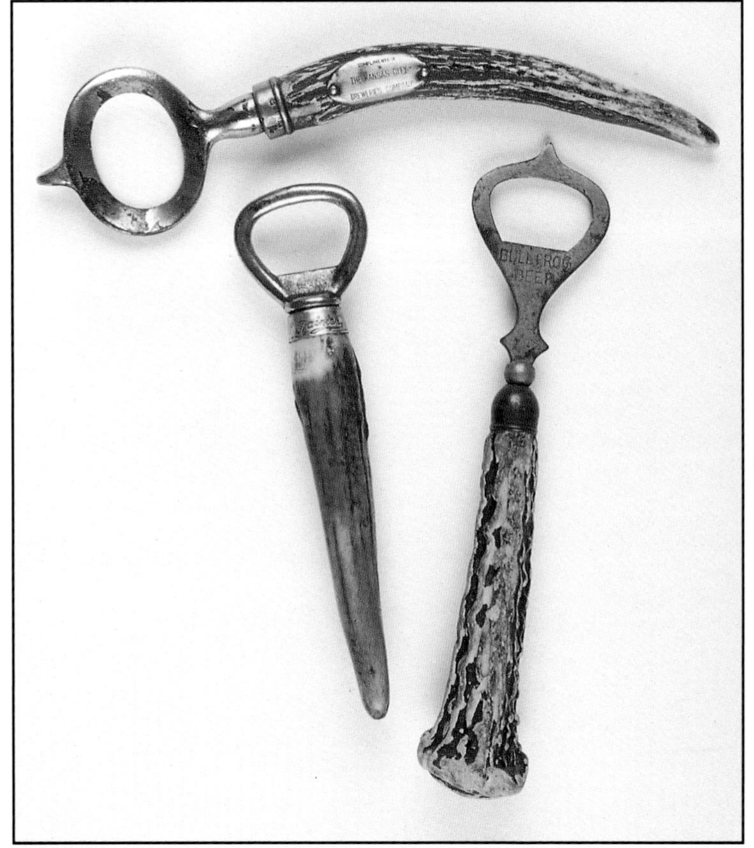

Top: M-52 Cap lifter with loop seal remover. Bone handle. Marked PATD 1906. Collar marked STERLING. American design patent 38,166 issued to John Hasselbring, August 14, 1906. 6 1/4". $75-100. *Advertising plate reads "Compliments of Kansas City Breweries Company."*
Bottom left: M-21 Cap lifter with bone handle. Various lengths. $25-35. *Found with collar marked RAINIER and STERLING.*
Bottom right: M-48 Cap lifter with loop seal remover. Bone Handle. Marked PAT'D FEB 6 '94. 6 1/2". $75-100.

Top left pair: M-23 Combination cap lifter, can piercer, and corkscrew. Made by EKCO, Chicago. Plastic handle. 5 1/4". $5-10.
Bottom left pair: M-39 "Tap Boy." Cap lifter, can piercer, and corkscrew. Made by Vaughan Company, Chicago. American design patent 170,999 for a "Bar Tool" issued to Michael J. LaForte, December 1, 1953. The tool was manufactured by Vaughan Company of Chicago under the name "Tap Boy." Usually marked VAUGHAN'S TAP BOY, PAT. NO. 170,999 , CHICAGO 24, U. S. A. 4 3/4". $10-15. *A souvenir example from Vaughan recognized the "U. S. Brewers Foundation Convention, New Orleans, 1952."*
Right: M-111 Cap lifter with corkscrew and single knife blade. Plastic handles. 3 3/4". $5-8.

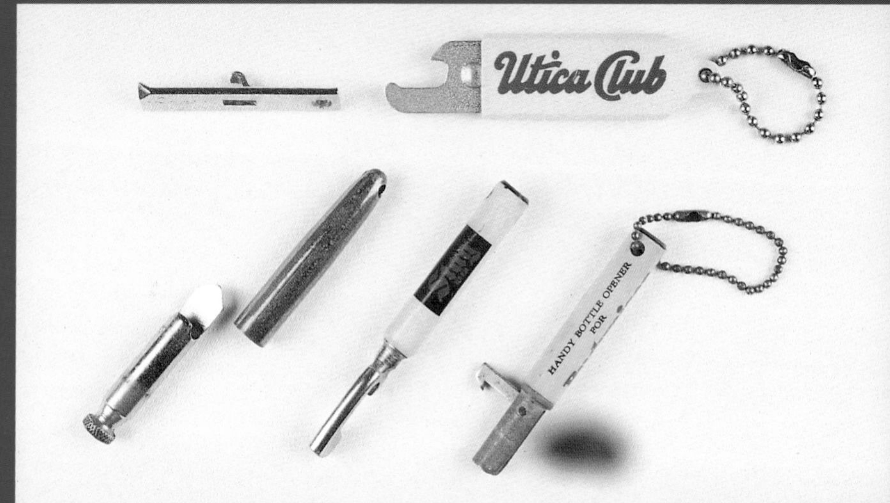

Top left: M-24 Opener for push tab cans. 2 1/8". $3-5.
Top right: M-49 Slide out cap lifter with plastic handle. 3" to 3 1/2". $5-10.
Bottom left: M-41 "Ke-Open-All" cap lifter and can piercer with metal sleeve. Made by Earl Products Co., Chicago, Illinois, and originally sold for as little as 10 cents. 3 1/8". $20-25.
Bottom middle: M-60 Cap lifter. Plastic handle. 3 1/4". $5-10.
Bottom right: M-89 Folding cap lifter. 2 7/8". $10-15.

Left: M-25 Nickel plated brass bullet roundlet with corkscrew (see L-2). 2 7/8". $25-35.
Middle pair: M-26 Pocket corkscrew in metal sleeve. Sleeve serves as a handle when in use. Marked WILLIAMSON CO. MANF'RS. NEWARK, N.J. 3 1/4". $75-100.
Right four: M-73 Slide out opener with folding corkscrew. 2 1/2" closed. 4 1/8" to 4 3/8" open. $75-125.

Left: M-28 Folding cap lifter and can piercer in leather case. Marked LATAMA ITALY. 4 1/4". $35-40.
Right: M-62 Sliding cap lifter and can piercer. Imitation bone handle. 4 3/4". $35-40.

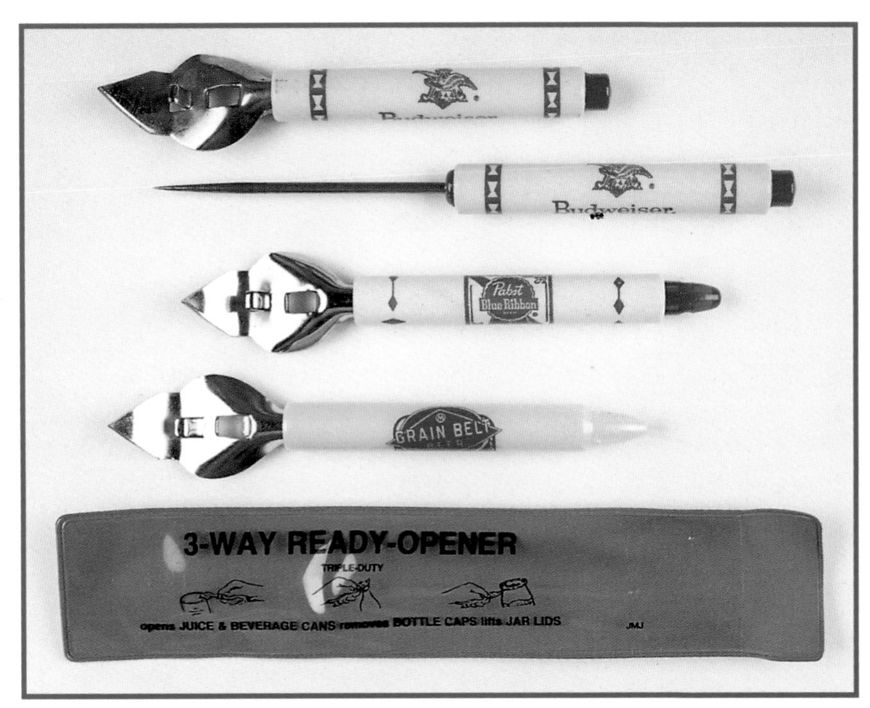

Top to bottom: M-29 Cap lifter and can piercer with plastic handle. Flat end. Developed by Mr. Lipic of St. Louis, Missouri, in the early 50s. 5 5/8". Shown with ice pick - supplied as an original set called "'Sportsman 3-way can opener, bottle opener, ice pick." $1-5. M-30 Cap lifter and can piercer with plastic handle. Pointed end. 5 3/4". $1-5. *This type is often used by distributors. They list the brand name(s) and their location creating a whole new area for the specialist collector. Locations on this type for Anheuser-Busch products distributed include: Jacksonville, Jacob, Decatur, and Waterloo, Illinois; St. Louis and Washington, Missouri; Hamburg, Iowa; Stratford, Wisconsin; Clearwater, Florida; and Roundup, Montana.* M-83 Cap lifter and can piercer. Plastic handle. Handle and end molded as one piece. 6 1/8". $1-5.

Top to bottom: M-31 Can piercer with plastic handle formed for finger grip. Some marked VAUGHAN CHICAGO MADE IN U. S. A. PAT. 1,996,550. 5 1/2". $5-10. M-34 Cap lifter with screwdriver and lid lifter. Marked ALEXANDER HUSKY, U. S. A. Made by Alexander Manufacturing Company, St. Louis, Missouri, beginning in 1975. 5 5/8". $5-10. *Tired of the same old beer? Here's an advertisement on this type that might help: "Falls City Beer. When you're ready for a bigger taste in beer."* M-38 Cap lifter and can piercer with plastic handle. Marked FOLEY. 5 3/4". $5-8. M-94 Cap lifter and can piercer. 4 1/4". $5-8.

Above: **Top:** M-36 Cap lifter with wood handle. Wire formed opener. 5 1/2". $20-25. **Middle:** M-57 Cast iron cap lifter with wood handle. 5 7/8". $40-50. **Bottom:** M-97 Cap lifter with wood handle. Marked ITALY PATENT. 5 5/8". $5-8. *One example is marked with the Anheuser-Busch Eagle in A trademark and MILITARY SALES.*

Left: **Left:** M-35 "Nite Club." Combination cap lifter, can piercer, jar top lifter, and muddler. 10". $10-20. *One advertising Storz Beer comes with a card: "The Storz Nite Club Opener is a valuable accessory for your bar or kitchen. 24-Carat Gold plated. Beer can opener, Bottle opener, Jar top lifter, cracks ice cubes, mixed drink muddler, crushes sugar cubes."* **Middle:** M-98 Fishing knife with opener at base of blade. Screwdriver tip at end of handle. Marked STAINLESS STEEL JAPAN. 12". $10-15. **Right:** M-104 Fishing knife with opener at base of blade. Cocktail fork at blade end. Screwdriver tip at end of handle. Marked STAINLESS STEEL JAPAN. 11 3/4". $10-15.

The type M-40 comes in a package describing it as "The Perfect way to open beer and soda bottles, also beer and juice cans. Precision made tool steel, with polished capehorn handle. Triple chromium plated for lifetime service. Langer Mfg. Co., New York 1, N. Y." Compare this type to M-50. The box for that one says "Langner Mfg. Co." Which is correct?

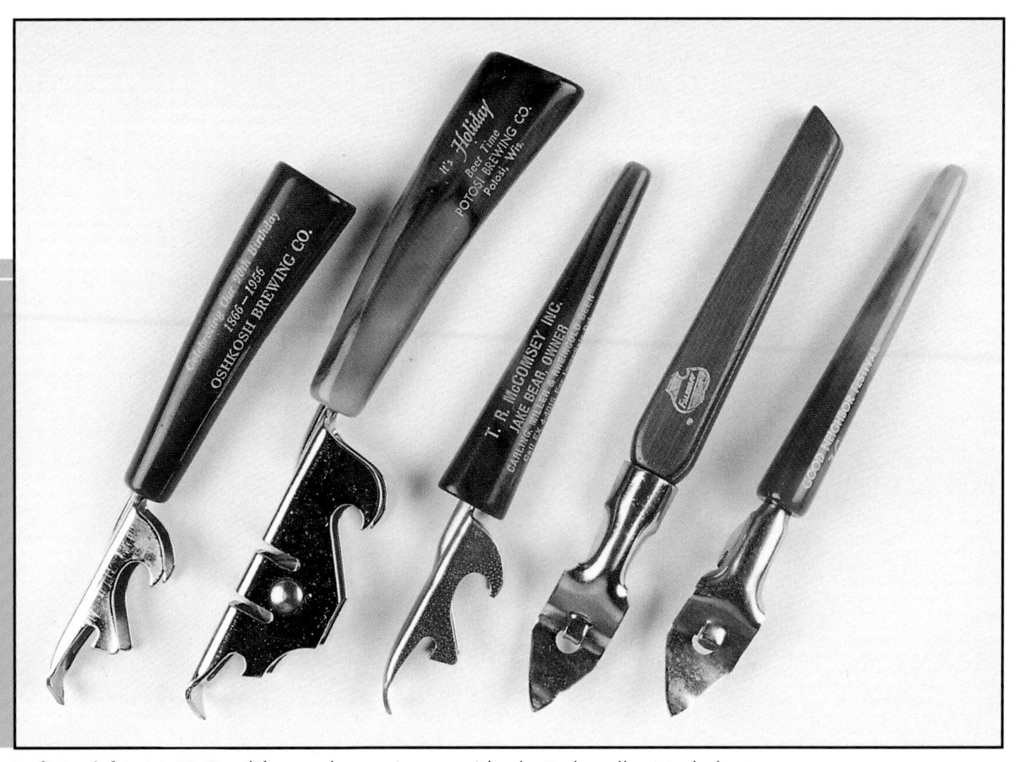

Left to right: M-40 Cap lifter and can piercer with plastic handle. Made by Langer Mfg. Co. of New York, New York. 5 3/4". $15-20. *One celebrated a birthday with: "Oshkosh Brewing Co., Celebrating our 90th Birthday, 1866-1956."* M-69 Cap lifter, can piercer, and knife sharpener. Plastic handle. 7 1/2". $35-40. M-96 Cap lifter and can piercer. Plastic handle. 6". $10-12. M-99 Cap lifter and can piercer. 6 5/8". $15-20. M-107 Cap lifter and can piercer. Plastic handle. 6 1/8". $10-12.

Above: **Left:** M-42 Cap lifter with plastic barrel. 4 1/4". $20-25. **Middle:** M-44 Cap lifter with graduated jar lid remover. 4 1/4". $10-15. **Right:** M-46 Cap lifter and can piercer with plastic bottle shaped sleeve. 5 3/4". $20-25.

Right: **Top, left to right:** M-43 Luggage tag cap lifter. Made in Denmark. 3 3/8". $1-5. M-81 Luggage tag cap lifter. 3 1/4". $1-5. M-82 Luggage tag cap lifter. 3 1/4". $1-5. M-90 Luggage tag cap lifter with twist off cap remover. 3 1/2". $1-5.
Bottom, left to right: M-91 Luggage tag cap lifter. Four supporting ribs. 3 1/4". $1-5. M-92 Luggage tag cap lifter. 3 3/4". $1-5. M-93 Luggage tag cap lifter. 4". $1-5. M-113 Luggage tag cap lifter. 3 1/2". $3-5. M-114 Luggage tag with twist off cap remover. 2 1/8". $3-5.

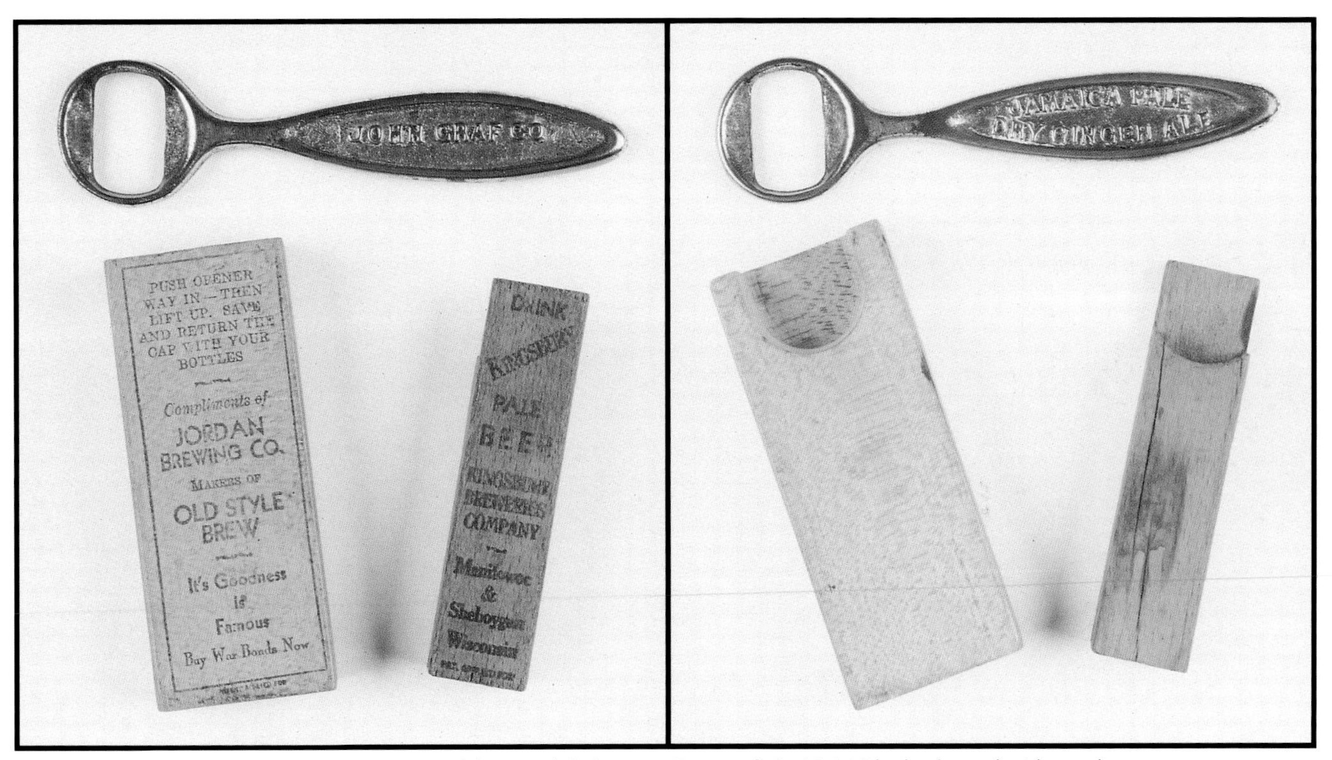

Top: M-45 Cast iron cap lifter. 5 1/4". $15-20. **Bottom left:** M-61 Block of wood with notch in the back for removing bottle caps marked PATENT APPLIED FOR. Reads "Push opener way in - then lift up. Save and return cap with your bottles. Compliments of Jordan Brewing Co., makers of Old Style Brew, It's [*sic*] goodness is famous. Buy war bonds now." 4 1/4". $35-40. **Bottom right:** M-80 All wood cap lifter marked PAT APPLIED FOR. 4". $35-40.

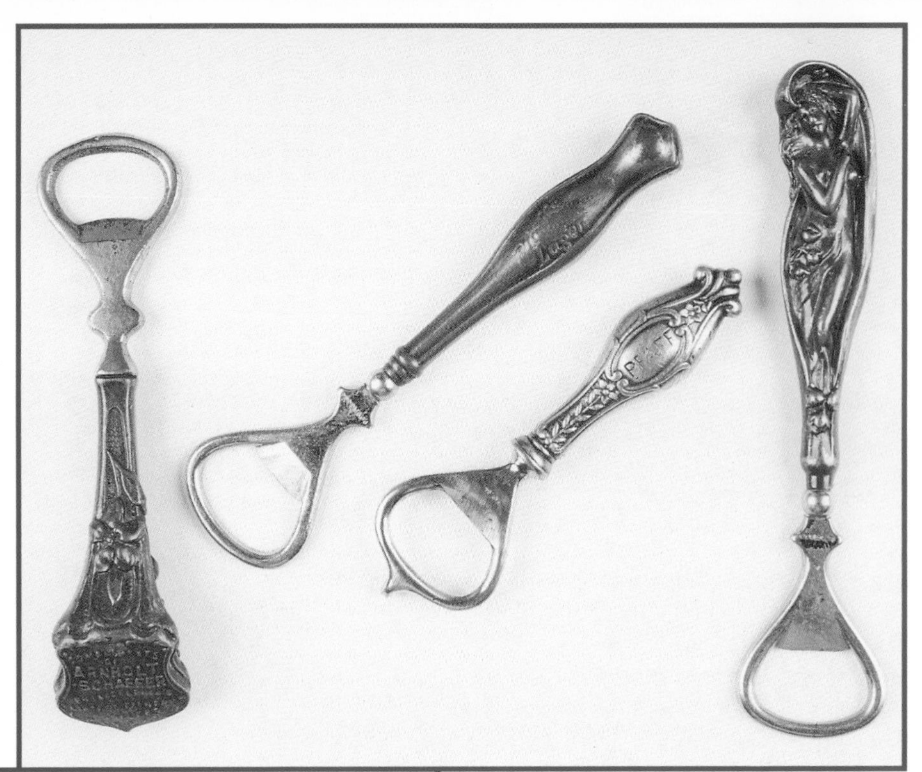

Left: M-47 Cap lifter with ornate cast handle. Marked PAT'D FEB 6 '94 (William Painter's patent). 5 5/8". $75-100. **Top middle:** M-56 Cap lifter with tableware quality handle. Marked PAT'D FEB 6 '94. 6". $60-75. *Found only with advertising: "Pfaff's Lager, 1857-1907."* **Bottom middle:** M-68 Cap lifter with loop seal remover. Marked PAT. 94. Made by Crown Cork & Seal Co. 4 5/8". $60-75. **Right:** M-103 Cap lifter. Ornate Art Deco handle with nude figure. Marked PAT FEB 6, 94. 6 1/2". $75-100.

Top: M-85 Cap lifter and can piercer. Plastic handle. Anheuser-Busch Eagle-In-A medallion in handle. Part of a set—see P-131. 7 3/8". $15-20. **Bottom:** M-50 Cap lifter and can piercer with plastic handle. Made by Langner Mfg. Co. of New York, New York. Packaging says "Four Way Opener. A must in every household. Opens those hard to open Vacuum Bottles - Ketchup Bottles and Jars of all kinds. Pierces Beer and Juice Cans. Removes Caps of Soda and Beer Bottles. Made of Tempered Tool Steel for Lifetime Service." 7 1/4". $15-20.

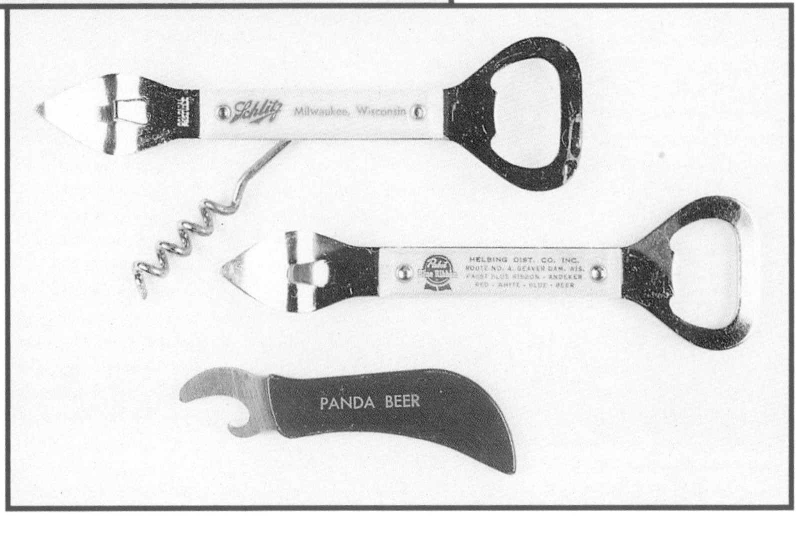

Top: M-53 Combination cap lifter, can piercer, and corkscrew. Plastic handle. Manufactured by Colonial Knife Company, Providence, Rhode Island. 5 5/8". $2-10. **Middle:** M-54 Cap lifter and can piercer. Plastic handle. Manufactured by Colonial Knife Company, Providence, Rhode Island. 5 5/8". $2-10. **Bottom:** M-74 Cap lifter with hand comfort formed plastic handle. 3 1/4". $10-15.

Left: M-55 Cap lifter and can piercer. Tap knob handle. 9 1/2". $25-30. **Right:** M-109 Two cap lifters, can piercer, serrated edge, screwdriver tip, lid punch. Wood handle. Marked LYMAN DOUBLE PUNCH LYMAN METAL PRODUCTS NORWALK, CONN. 7 1/8". $35-45.

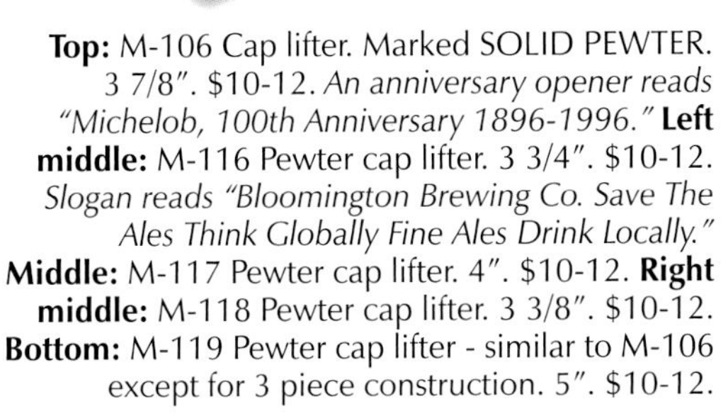

Top: M-106 Cap lifter. Marked SOLID PEWTER. 3 7/8". $10-12. *An anniversary opener reads "Michelob, 100th Anniversary 1896-1996."* **Left middle:** M-116 Pewter cap lifter. 3 3/4". $10-12. *Slogan reads "Bloomington Brewing Co. Save The Ales Think Globally Fine Ales Drink Locally."* **Middle:** M-117 Pewter cap lifter. 4". $10-12. **Right middle:** M-118 Pewter cap lifter. 3 3/8". $10-12. **Bottom:** M-119 Pewter cap lifter - similar to M-106 except for 3 piece construction. 5". $10-12.

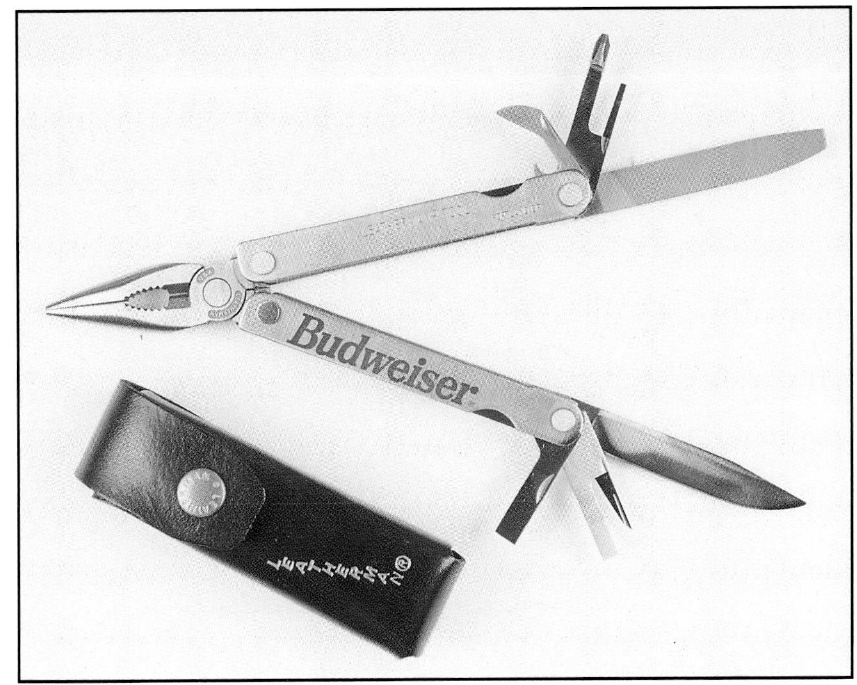

M-110 Leatherman pocket survival tool. Includes can-bottle opener, file, Phillips screwdriver, pliers, needle nose pliers, wire cutters, knife, nail nick, awl, and flat small, medium, and large screwdrivers. Marked LEATHERMAN TOOL PORTLAND OR. 4" folded. $40-50.

Top: M-63 Painted aluminum cap lifter. 10". $15-25. *One advertises "Sebewaing Brewing Co., Sebewaing, Michigan since 1886, celebrating our diamond anniversary" (75 years = 1961).* **Bottom:** M-112 Cap lifter. Plastic on steel. 3". $10-12.

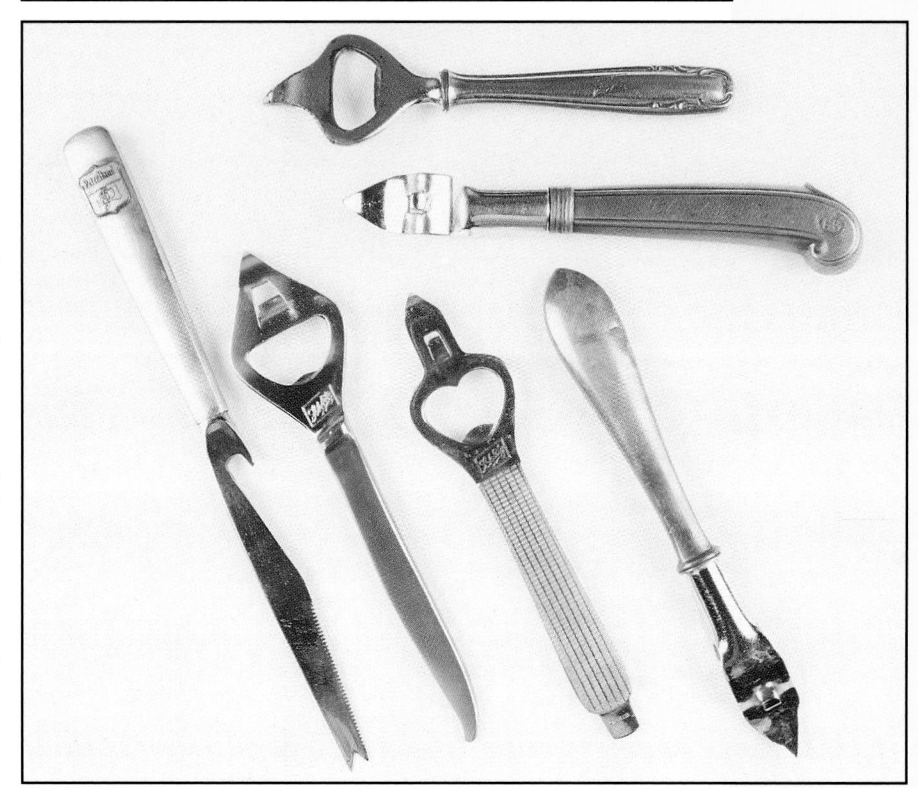

Top right: M-66 Cap lifter and can piercer. Tableware handle marked STAINLESS, GERMANY. 6 1/8". $20-25. **Second from top right:** M-95 Cap lifter. Tableware handle marked STERLING HANDLE. 6 1/4". $40-50. *Simple advertising has a "GB" shield logo (Griesedieck Brothers) and "Christmas '53."* **Bottom, left to right:** M-75 Cap lifter at base of fixed blade knife. Tableware handle marked STEGOR STAINLESS. 9". $40-50. M-76 Cap lifter and can piercer. Tableware handle marked STAINLESS AUSTRIA. 7 1/4". $30-35. M-86 Cap lifter and can piercer. Marked LUNAWERK GERMANY. Leather pouch. 6 1/4". $30-35. *The opener celebrates "Falstaff, 1964 Falstaff NBWA, 'The World's Fair City,' New York." The pouch sports the Falstaff shield logo.* M-88 Cap lifter and can piercer. Tableware handle with opener marked STERLING HANDLE W (in logo) PAT.PENDING. 6 7/8". $40-50.

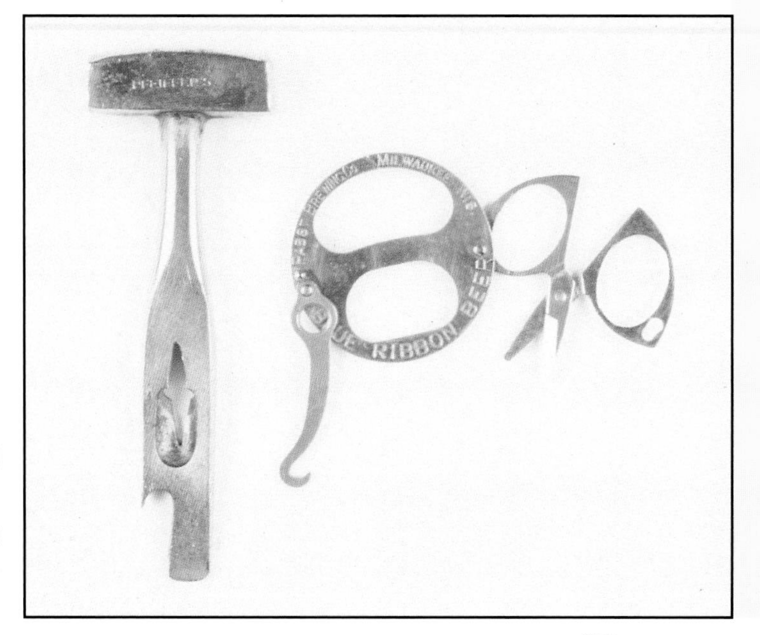

Left: M-77 Cap lifter, ice cracker, hammer, screwdriver, nail remover, and cigar tool. Marked PAT. MCH. 5, 1901. 4 3/4". $60-75. **Right:** M-78 Cap lifter, button hook, scissors, key chain, and cigar cutter. Marked PAT. APRIL 14, 1914 HESUAH. 1 7/8". $75-100.

Top left: M-79 Cap lifter and can piercer. 3 3/4". $10-15. **Top middle:** M-105 Cap lifter. 5 1/2". $5-8. **Top right:** M-115 Cap lifter with magnetic back. 3 1/4". $3-5. **Bottom:** M-87 Cap lifter and can piercer. 9 5/8". $10-15.

Right & below: **Top:** M-100 Cap lifter. Plastic handle. Two sizes of plastic bottle recappers on reverse. Marked PAT.PENDING. 5 3/4". $10-15. **Bottom:** M-108 Wire cap lifter with three frets. Plastic handle. Plastic bottle capper is held in end opposite opener. Marked B&B REMEMBRANCE ST. PAUL, MINN. U.S.A. on opener end and on plastic bottle cap. 5 3/4". $20-25.

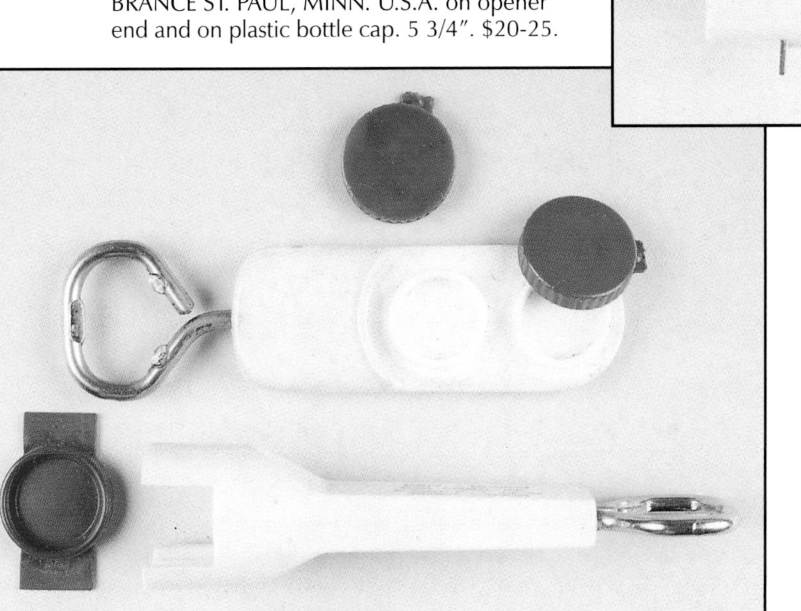

Novelty Openers

A wide variety of openers fall under this category which, generally, includes those openers that have accessories to serve other purposes. There are knives, cigar cutters, shoe horns, pencils, jar openers, jiggers, button hooks, money clips, and bottle sealers and resealers. And we've even thrown in a fishing lure, a baseball bat, and a belaying pin.

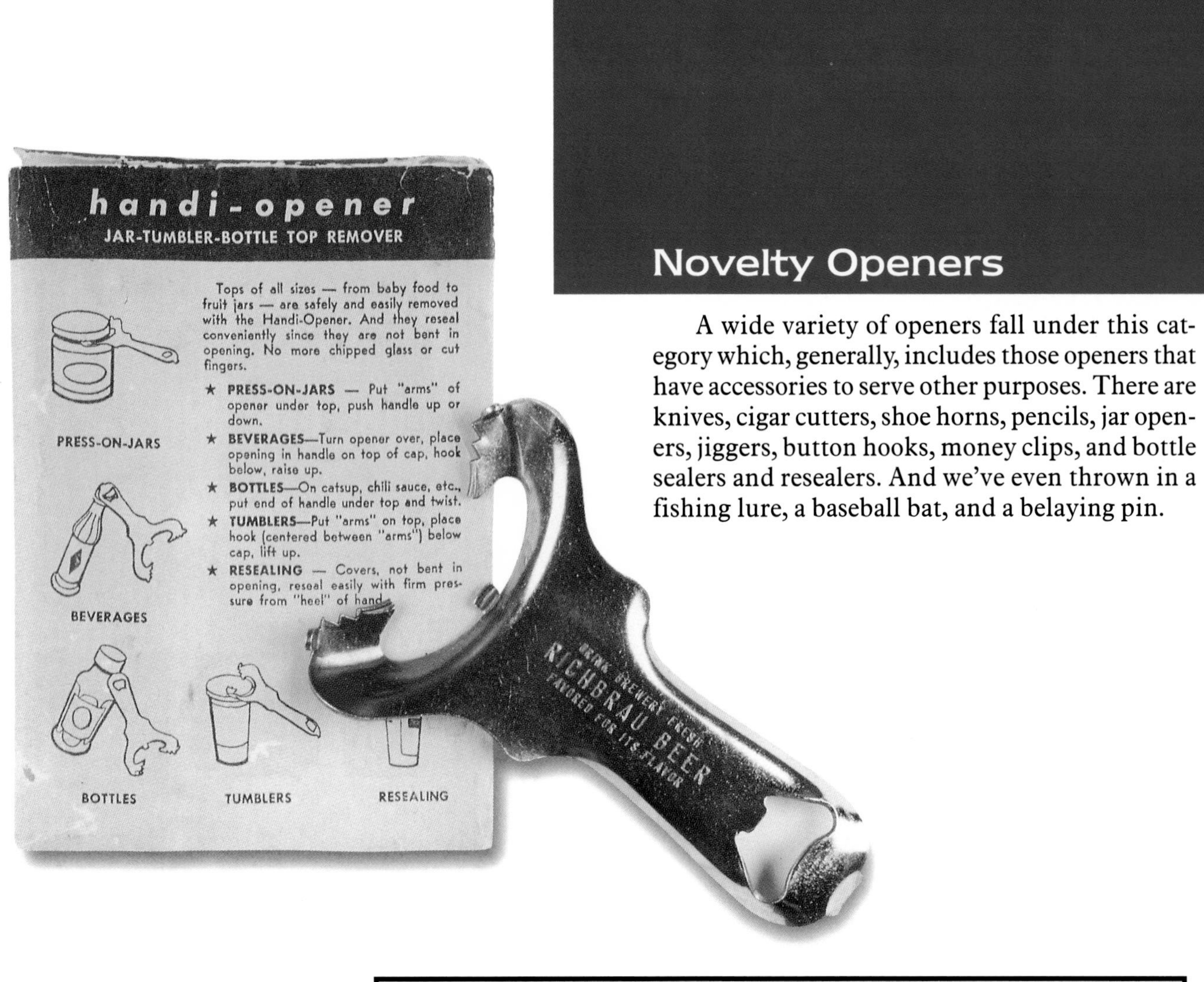

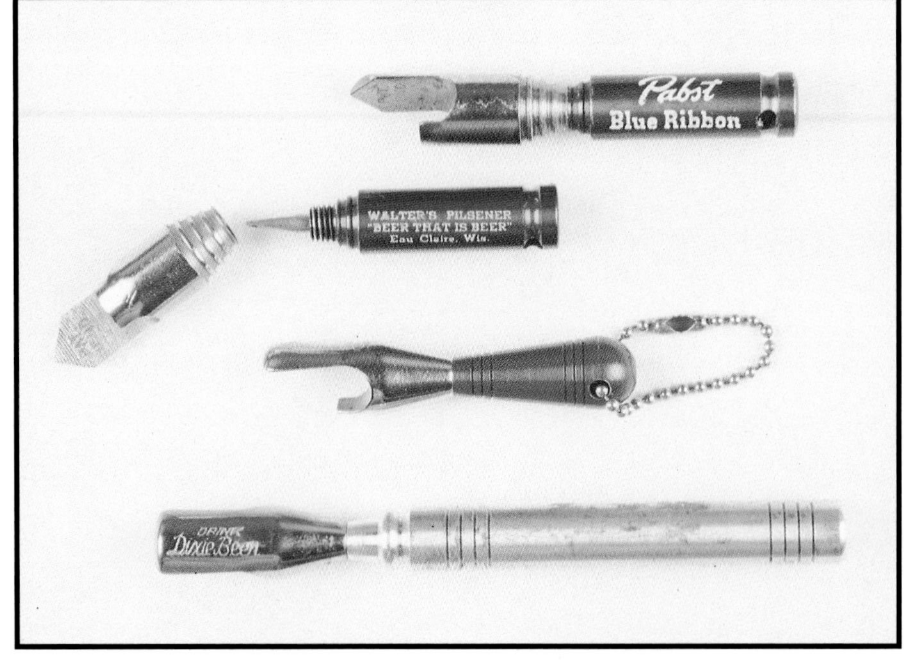

Top to bottom: (pair) N-1 Cap lifter with screwdriver. c.1939. 3". $40-50. N-2 Cap lifter with bowling pin shape handle. American design patent 153,349 issued to Oscar Galter, April 12, 1949. 2 1/2". $10-12. N-88 Cap lifter with cylindrical handle. Supposedly local bartenders in the New Orleans area were not happy with type N-2 because the handle is so short. To satisfy their demands, type N-88 was developed with a much longer handle. 4 3/4". $15-20.

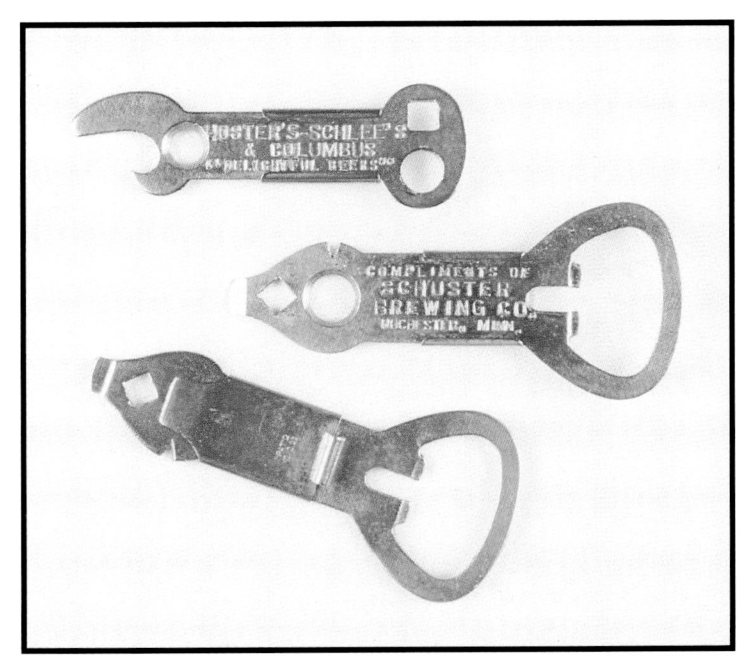

Top: N-7 Cap lifter with sliding cigar cutter and Prest-O-Lite key. Marked PAT. 10.12.09. American patent 936,678 issued to John L. Sommer, October 12, 1909. 3". $100-125. **Bottom pair:** N-8 Cap lifter with sliding cigar cutter, cigar box opener, nail puller, screw-driver tip, and Prest-O-Lite key. Marked PAT. 10.12.09. American patent 936,678 issued to John L. Sommer, October 12, 1909. 3". Obverse and reverse shown. $75-100.

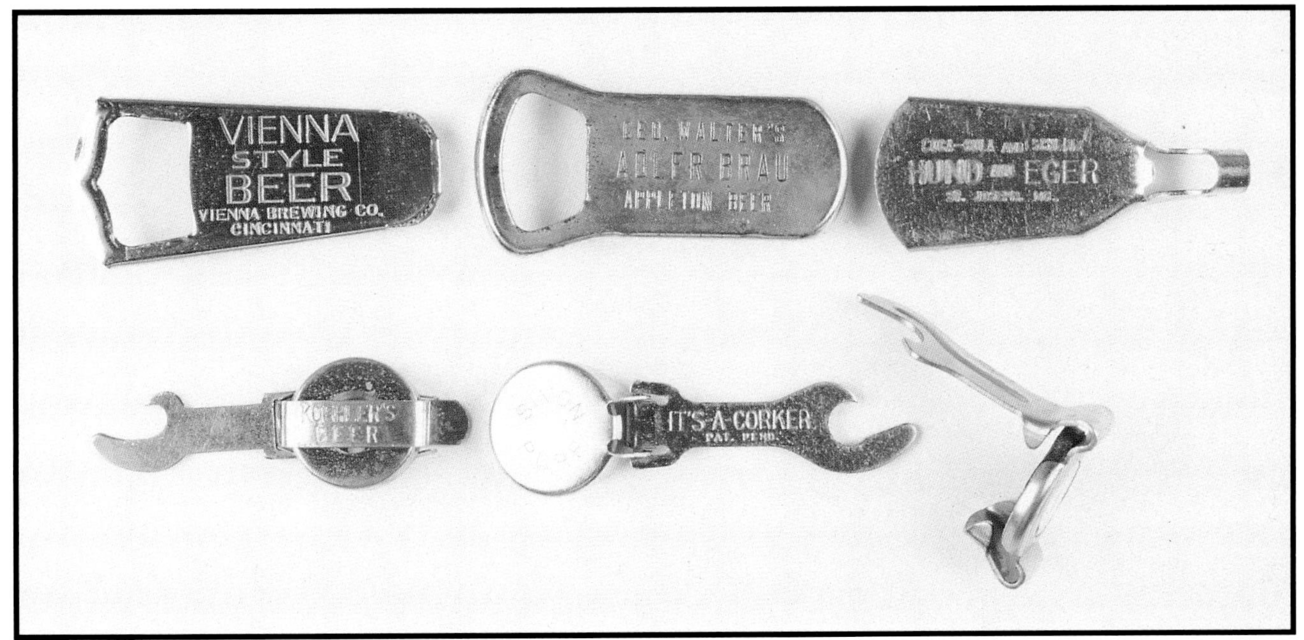

Above: **Top, left to right:** N-9 Cap lifter and bottle resealer. Appears in a 1937 Vaughan Novelty Manufacturing catalog as "Locktite (Trade Mark) Bottle Stopper and Opener Combination." It is described as "Removes the crown caps. The rubber gasket seals bottle air tight, preserving the unused portion for future use. Will last for years. Large space for advertisement." 3 1/8". $10-30. N-83 Cap lifter with bottle resealer. Swinging square wire on underside. Marked PAT. PEND DEPENDABLE MILWAUKEE. 3 1/4". $40-50. N-62 Cap lifter with bottle resealer. Marked PAT. PEND., B & B, ST. PAUL (Brown & Bigelow Company). American design patent 98,486 issued to M. E. Trollen, February 4, 1936. 3 1/4". $40-60. **Bottom, left to right:** N-10 Cap lifter with bottle stopper. Marked PAT'D U. S. A. DEC. 9, 19, NOS. 1,324,256. American patent 1,324,256 issued to William B. Langan, December 9, 1919. Patent assigned to the Koscherak Siphon Bottle Works, Hoboken, New Jersey. 3 1/8". $10-25. N-79 Cap lifter with bottle stopper. Marked IT'S-A-CORKER PAT. PEND. 3 5/8". $25-30. N-58 Cap lifter with bottle resealer. Marked SAV-KAP BOSTON PAT. DEC. 7, 1926. 2 3/8". $30-35.

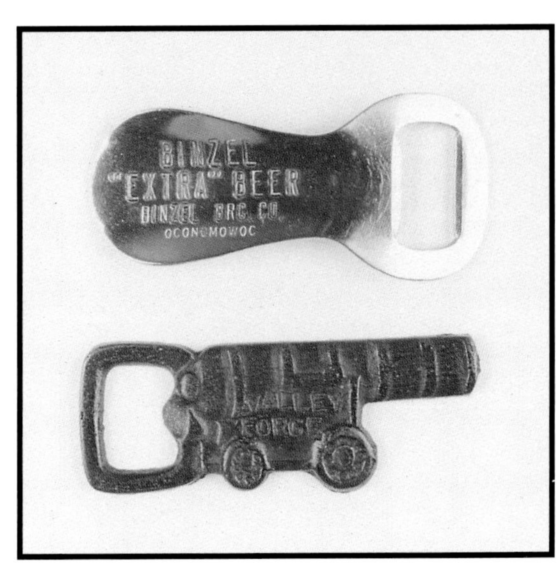

Right: **Top:** N-11 Cap lifter on shoe horn. Marked PAT APL. FOR. Shown in a 1961 Handy Walden catalog. A slightly different design is shown in a 1971 Vaughan catalog. 3 3/8". $10-25. **Bottom:** N-16 Cast iron miniature cannon. 3 1/2". $10-15.

Top to bottom: N-12 Cap lifter and lighter. Remove round end cap to expose lighter. Marked REDILITE, PAT. 1,820,131, MADE IN U. S. A., B. & B., ST. PAUL, MINN. on the lighter end. American patent 1,820,131 issued for the lighter to Howard L. Fischer, August 25, 1931. 6 1/2". $40-50. **(pair)** N-22 Cap lifter with pencil. Marked AMERICAN PENCIL CO. NEW YORK PAT. PEND. U. S. A. KAPOFF OPENUP. 7" (length will be less if pencil sharpened!). $30-50. N-67 Mechanical pencil with cap lifter. 5 1/2". $40-50.

Recipe for a
CONTENTMENT
COCKTAIL

TAKE one lazy evening and garnish liberally with congenial companions.

Add a dash of high spirits and mix according to taste and inclination.

Light up your favorite smoke with your new bar-tool Redilite, whap off the caps from a few bottles of BECK'S BEER with your bar-tool bottle opener and settle down to enjoying yourself.

MAGNUS BECK BREWING COMPANY, INC.
Brewing the BEST, Not the Most
Since 1855

Instructions for operation and care of

The "Redilite" Pocket Lighter

TO LIGHT—Remove cap A and with thumb give abrasive wheel a quick sharp turn.

TO EXTINGUISH—Replace cap.

TO FILL—Remove cap and holding the lighter near the base with one hand, withdraw the body of the lighter C by grasping it at the shoulder—just below the wheel—and pulling it straight out from the outer shell B.

Saturate the cotton in this inner shell with small quantity naphtha, benzine, energine or any of the standard lighting fluids and shake out the excess fluid.

DO NOT FLOOD CONTAINER. USE JUST ENOUGH LIQUID TO BE HELD IN SATURATION LEAVING NO EXCESS OR FLOWING LIQUID TO RUN OUT ON THE LIGHTER.

TO ADJUST FLINT—Remove cap and withdraw body as in filling. Turn flint adjusting screw D clockwise until flint is properly raised against abrasive wheel at top.

TO INSERT NEW FLINT—Unscrew flint adjusting screw and draw out tension spring and plunger. Shake out worn flint, drop new one in tube and reassemble, turning adjusting screw until the flint is in position to give good spark as wheel is turned.

WICK—The wick is asbestos and will last indefinitely.

This Redilite has been carefully inspected and tested and leaves our factory in perfect condition.

It is guaranteed by "The House of Quality" to give satisfaction.

BROWN & BIGELOW
ST. PAUL, MINN.

MADE IN U. S. A.

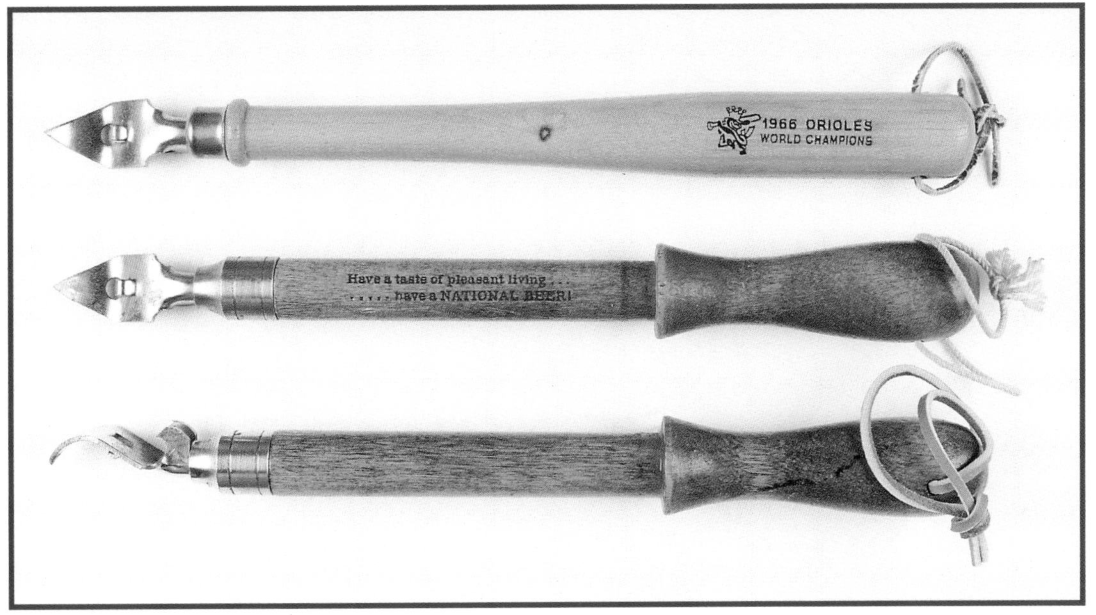

Top: N-13 Cap lifter and can piercer. Wood baseball bat shape. 12". $30-50. *A dated example advertises "National beer, 1966 Orioles World Champions."* **Bottom pair:** N-17 Cap lifter and can piercer. Belaying pin (used on Clipper ships to hang and release ships' lines and ropes). Designed by John Schneider of National Brewing Company, Maryland. Approximately 100,000 openers sold between 1960 and 1963. 9 1/2". $30-40.

Top: N-19 Bartender's tool with cap lifter, corkscrew, jigger, ice cracker, and drink recipe viewer. 6 1/2". $40-50. **Bottom:** N-64 "Bar Boy." A "six-in-one appliance" from Tempro Incorporated, New Haven, Connecticut. Including cap lifter, folding corkscrew, ice masher, and jigger. 7". $75-100.

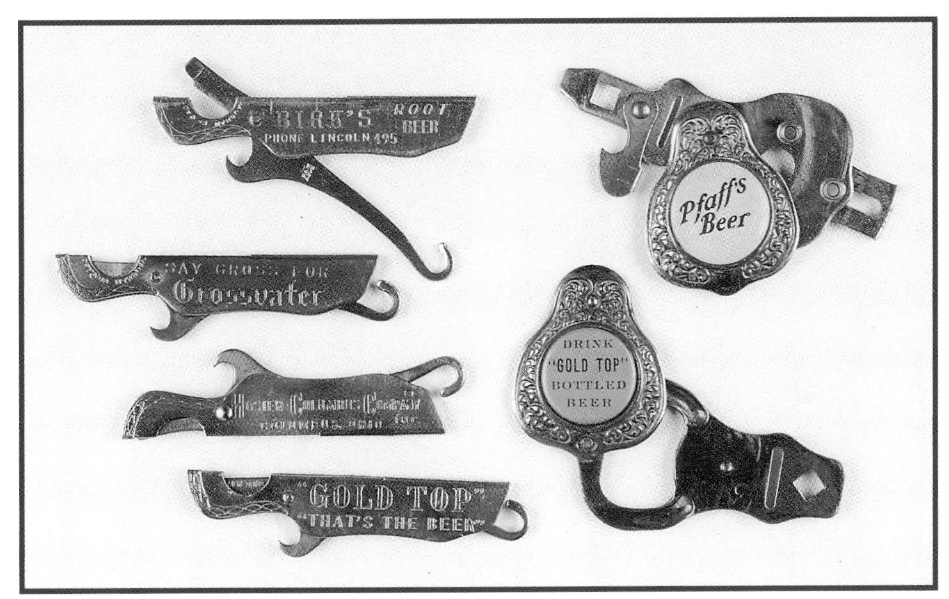

Left four: N-20 "Jim Dandy." A 4-in-1 pocket tool by Vaughan Company with bottle opener, button hook, cigar cutter, and screwdriver. Shown in a 1922 Vaughan catalog. 3". $25-40. **Top right:** N-49 Cap lifter, screwdriver, cigar cutter, Prest-O-Lite key, and watch fob. Marked PATENTED OCT. 8, '12 & OCT. 7, '13. Patented by Arthur Merrill. 3 1/8". $100-125. **Bottom right:** N-78 Cap lifter with Prest-O-Lite key. Watch fob. 2 1/2". $100-125.

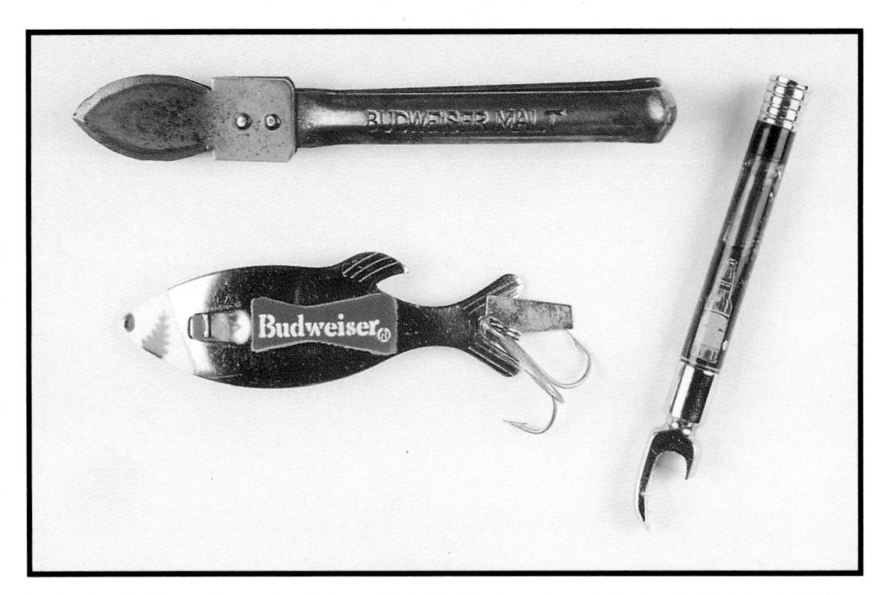

Top left: N-21 Opener for malt cans. Marked PAT.APPLD. 5 1/2". $40-50. *Advertises "Anheuser Busch, Budweiser Malt."* **Bottom left:** N-32 Cap lifter and can piercer on a fishing lure. Made by Heddon Company. American patent 2,986,812 issued to William Arter, Jr. and Robert J. Clouthier, June 6, 1961. Supplied in a box labeled "The Growler." 3 3/4". $30-40. **Right:** N-76 Cap lifter with bottle and glass of beer depicted on handle. When held one way, the glass fills up. When tilted in the opposite direction, glass disappears. 4 3/4". $40-50.

Top: N-26 Cap lifter, lid prier, and jar top opener. Marked B&B ST. PAUL, MINN. U. S. A. REMEMBRANCE PATENT 2,493,438 2,501,204, 2,501,205. American design patent 176,518 issued to James L. Hvale, January 3, 1956. 5 1/2". $25-35. **Bottom:** N-91 Shovel shape cap lifter with ice scoop. Marked JAPAN. 7 1/8". $20-25.

N-40 Belt buckle with cap lifter on reverse side, in the middle or at the other end. Some made by Bergamot Brass Works. American design patent 246,552 issued to Daniel Baughman, December 6, 1977. Various diameters and lengths. $10-25.

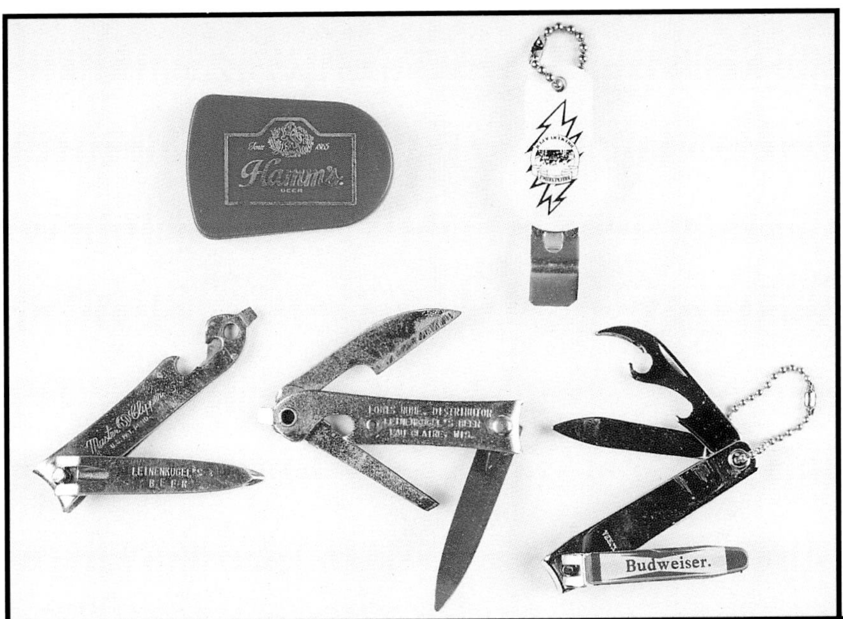

Top left: N-31 "Snap a Cap." Cap lifter and bottle sealer. Made by Teraco, Inc., Midland, Texas. Over one half million were produced from 1968 to 1974. 2 3/8". $10-12. **Top right:** N-30 Cap lifter and bottle sealer. Cap lifter slides into plastic sleeve. 2 3/8". $1-5. *One example sports a well-known early popular slogan "Don't let your spirits get low, say Hey Mabel, Black Label!"* **Bottom left:** N-38 Nail clipper/file with cap lifter and screwdriver on handle. Marked MASTER GSI (IN CIRCLE) CLIPPER, U. S. PAT. PEND. 3 1/8". $20-25. **Bottom middle:** N-57 "8 in 1" clipper. Combination tool with cap lifter. Marked GSI CLIPPER. 3 1/8". $20-25. **Bottom right:** N-65 Nail clippers with double cap lifter blade and master blade. Marked BELL. 2 3/4". $10-15.

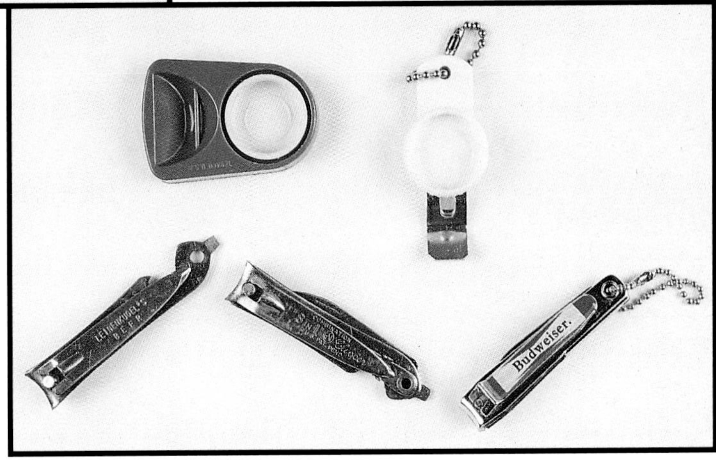

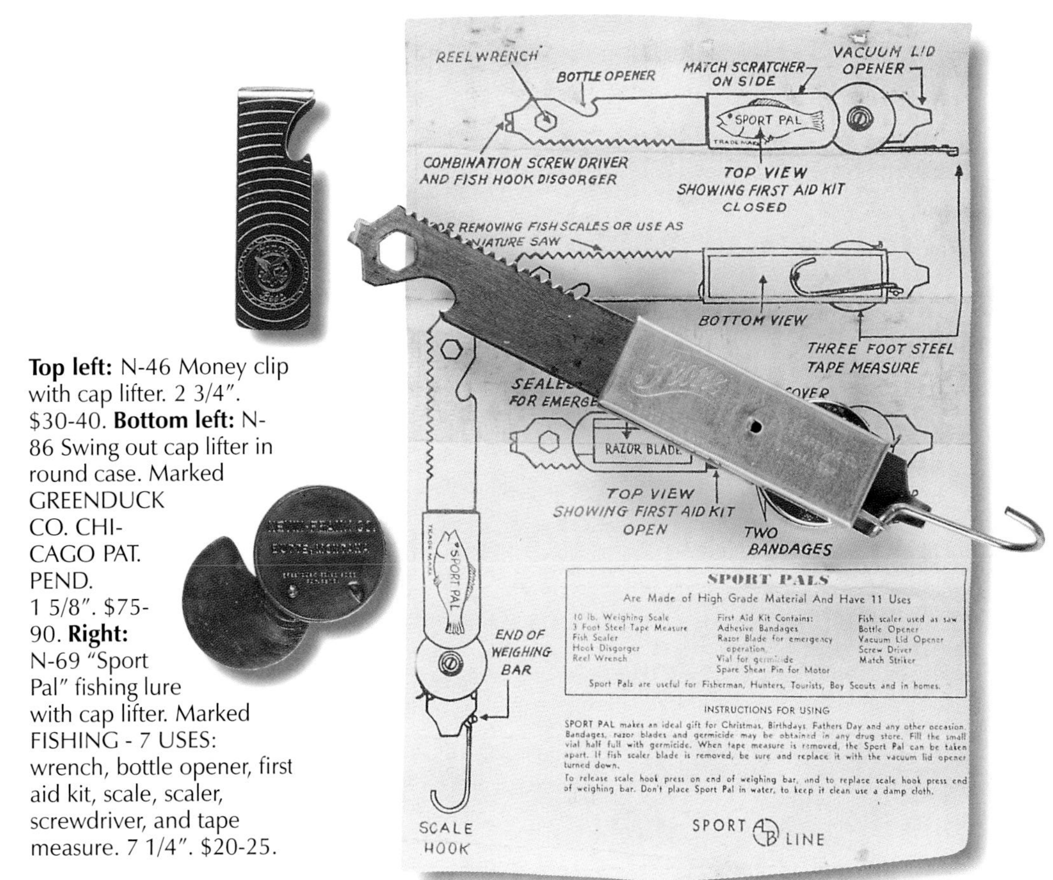

Top left: N-46 Money clip with cap lifter. 2 3/4". $30-40. **Bottom left:** N-86 Swing out cap lifter in round case. Marked GREENDUCK CO. CHICAGO PAT. PEND. 1 5/8". $75-90. **Right:** N-69 "Sport Pal" fishing lure with cap lifter. Marked FISHING - 7 USES: wrench, bottle opener, first aid kit, scale, scaler, screwdriver, and tape measure. 7 1/4". $20-25.

Instructions for using the N-69 Sport Pal:

Sport Pal makes an ideal gift for Christmas, Birthdays, Fathers Day and other occasion. Bandages, razor blades and germicide may be obtained in any drugstore. Fill the small vial half full with germicide. When tape measure is removed, the Sport Pal can be taken apart. If fish scaler blade is removed, be sure and replace it with the vacuum lid opener turned down.

To release scale hook press on end of weighing bar, and, to replace scale hook press end of weighing bar. Don't place Sport Pal in water, to keep it clean use a damp cloth.

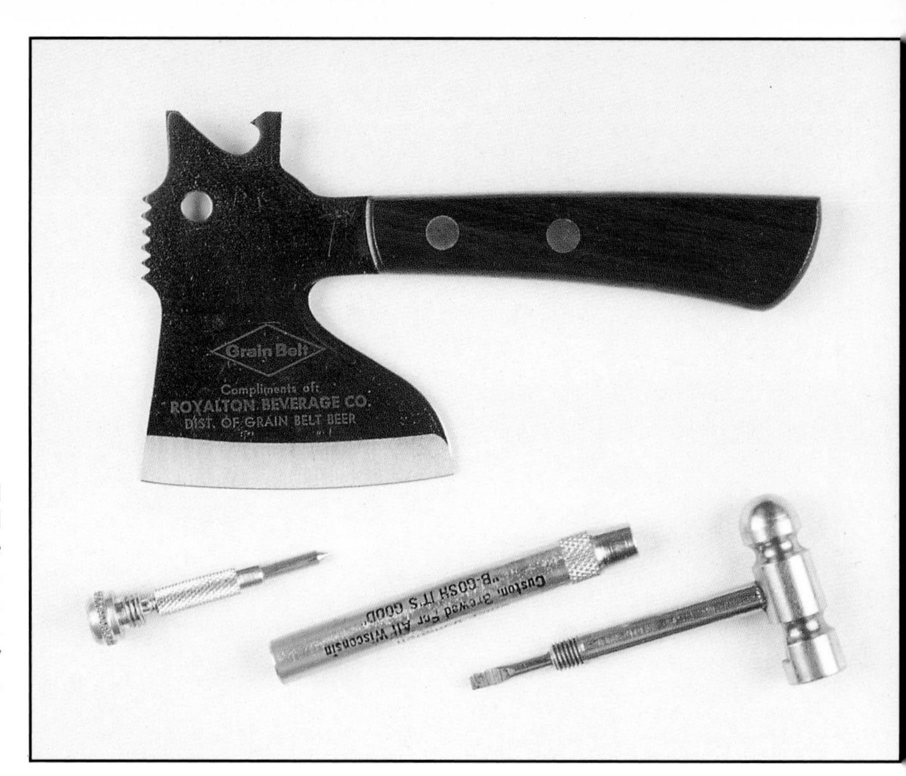

Top: N-71 Cap lifter with meat cleaver and meat tenderizer. Supplied in box labeled "Burnco 'George Washington's' Cheese Hatchet." Marked STAINLESS STEEL JAPAN. 5 1/2". $35-45. **Bottom:** N-77 Cap lifter, hammer, nut cracker, ice breaker, ball peen, awl or punch, and medium and tiny screwdrivers. 5 1/4". $25-35.

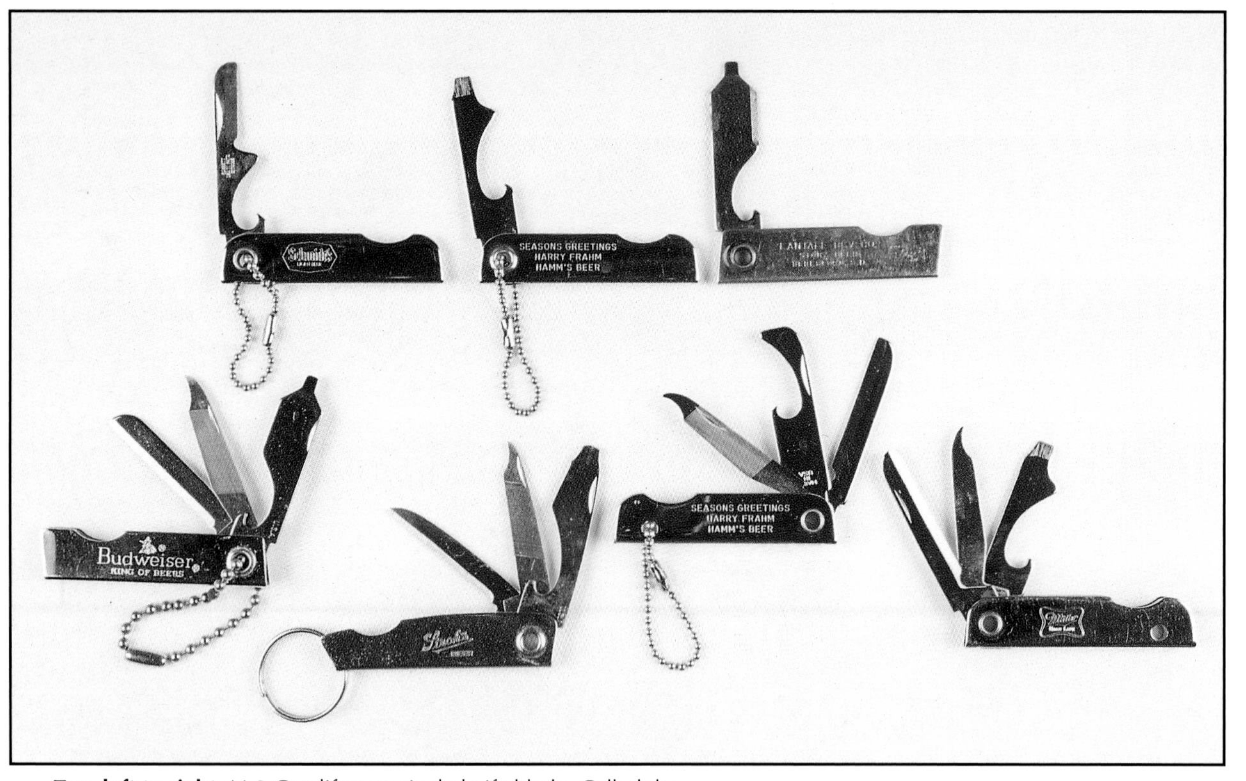

Top, left to right: N-3 Cap lifter on single knife blade. Called the "Derby Duke." Marked BASSETT U. S. A. PATD, 2,779,098. American patent 2,779,098 issued to Edward J. Pocoski and William G. Hennessy, January 1, 1957. 2 1/4". $3-15. N-75 Folding cap lifter with screwdriver tip folding into knife style handle. Marked BASSETT, MADE IN U. S. A., PATENTED. 2 1/4". $10-15. N-84 Knife with cap lifter/screwdriver/cutting blade. Marked U. S. A. 2 3/8". $10-15. **Bottom, left to right:** N-18 Three blade knife with cap lifter on one blade with screwdriver tip. 2 3/8". $10-15. N-37 Knife with pen blade, file blade, and cap lifter/screwdriver blade. Key ring attached. 2 1/2". $10-12. N-41 Knife with cap lifter blade, file blade, and pen blade. 2 3/8". $10-15. N-56 Knife with three blades including cap lifter. 2 3/8". $10-15.

N-95 Folding cap lifter/can piercer knife with one master blade. Marked PROV. CUT. CO. PROV. R.I. U.S.A. 3". $35-40.

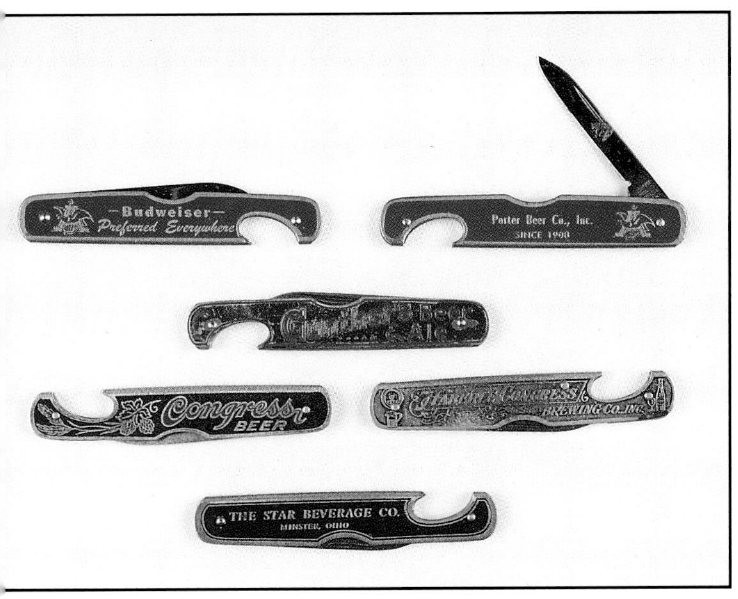

N-4 Cap lifter built into handle of single blade knife. Marked ETCHED P. CO., L. I. C., N. Y." (L. I. C. is Long Island City) or G. SCHRADE BPORT, CT. (Bridgeport, Connecticut). 3". $30-60.

Above: **Left pair:** N-5 Cap lifter on bolster of two blade knife. Boot shape with ivory handles. Marked UTICA CUTLERY, UTICA, N. Y. 3 1/4". $75-90. **Middle:** N-82 Boot shape single blade knife with cap lifter formed in handle. Celluloid handle. Marked SYRACUSE KNIFE COMPANY. 3 1/4". $35-75. **Right pair:** N-85 Boot shape single blade knife with cap lifter formed in metal handles. Marked KUTMASTER UTICA, N. Y. MADE IN U. S. A. 3 1/4". $60-75.

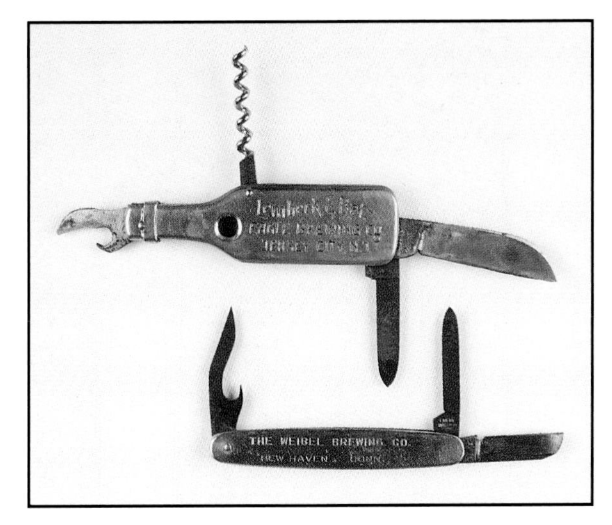

Right: **Top:** N-6 Cap lifter, corkscrew, cigar cutter, file blade, and single knife blade. Bottle shape. Marked GRIFFON CUTLERY WORKS, GERMANY. 3 1/4". $150-200. **Bottom:** N-33 Knife with cap lifter blade and master blade. Made by Empire, Winsted,

Top: N-15 Cap lifter with two blade knife. Shape of Budweiser "Bow-Tie." Marked MADE IN JAPAN. 2 5/8". $20-25. **Middle:** N-23 Eighteen wheel truck with cap lifter/screwdriver blade and single knife blade. Shown in the Colonial Knife Company, Providence, Rhode Island, February 1980 catalog. Prices ranged from $6.25 each for 100 to $4.90 for 2000. 3". $10-12. **Bottom:** N-25 Cap lifter with single blade knife handle. Blade marked COLONIAL PROV. U. S. A. 3 1/4". $10-12.

Top: N-24 Cap lifter formed in handle of single blade knife. Marked STAINLESS, SHEFFIELD, ENGLAND. 3". $15-30. **Middle:** N-34 Single blade knife with cap lifter formed in handle. Marked LATAMA, ITALY. 2 5/8". $30-40. **Bottom:** N-52 Single blade knife with cap lifter formed in handle. Made by Miller Brothers, Meriden, Connecticut. 3". $75-100. *A dated example reads "Emmerling Brewing Co., Johnstown 1913, Grossvader Beer."*

Top: N-36 "Fish-Knife" with cap lifter on scaler blade. Marked COLONIAL, PAT. NO. 2,310,641, U. S. A. 5". $40-50. **Middle:** N-42 Fishing knife with cap lifter and folding can piercer. Master blade marked P2170537 & 2281712, IMPERIAL PROV. U. S. A. Fish scaler blade marked BEVERAGE CAN OPENER 2361-889. 4 1/2". $40-50. **Bottom:** N-66 "Panther Angler" knife with master blade and cap lifter on fish scaler blade. Marked C.I. HI-STAINLESS 533 JAPAN on master blade. Box labeled "Compass Instrument & Optical Co., Inc." 4 1/2". $40-50.

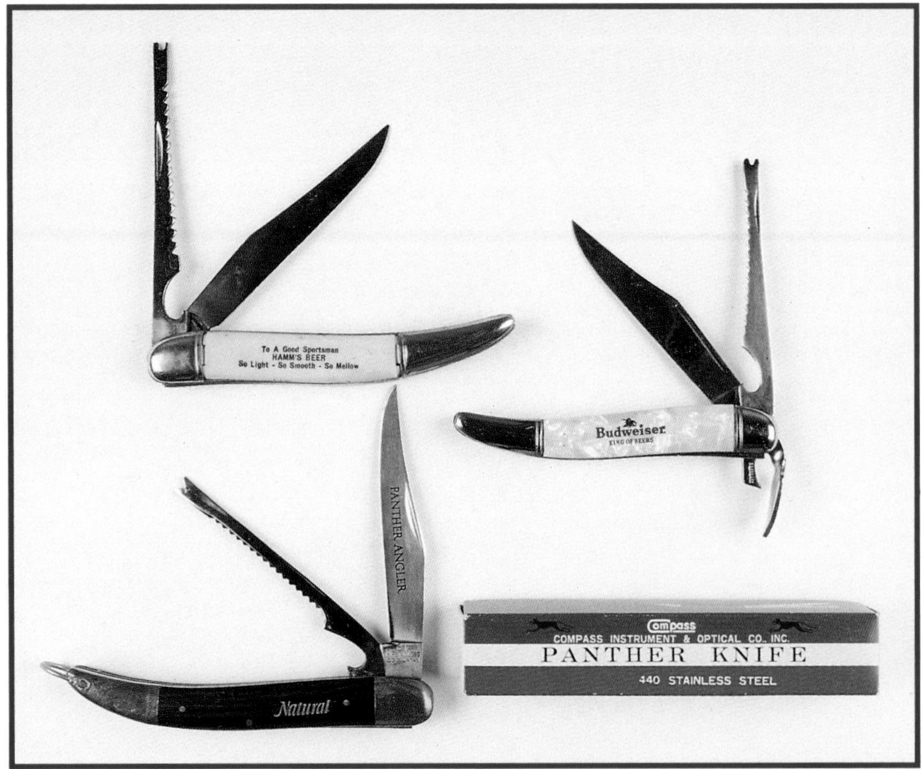

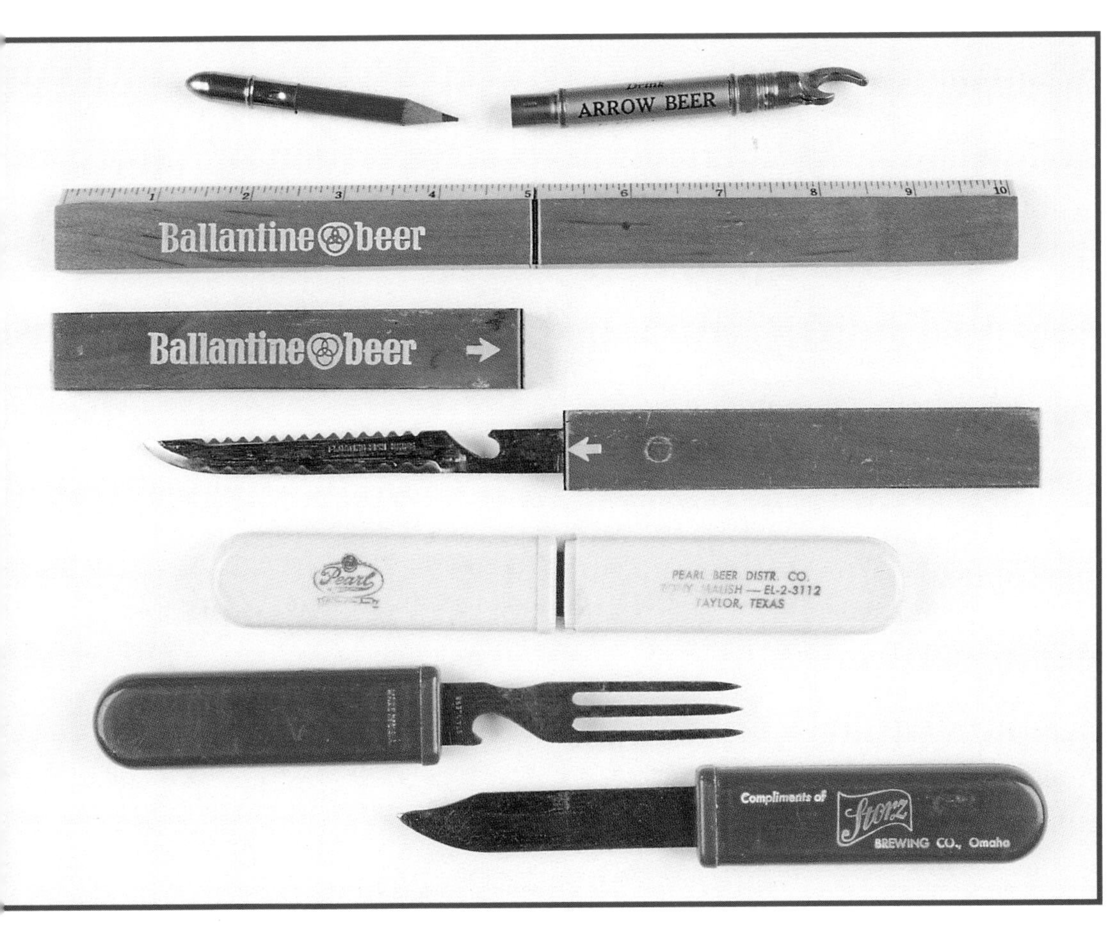

Top: N-29 Cap lifter. Bullet end contains a pencil which reverses and is inserted into the handle for use. Marked G. FELSENTHAL & SONS, CHICAGO. 4 5/8". $20-50.
Middle pair: N-27 Cap lifter on concealed blade of floating fish knife. Marked WARCO, STAINLESS STEEL, JAPAN. 10". Produced with and without ruler on the edge. $15-25. *An appropriate gift is stamped "To a good sportsman from the brewers of Holiday Beer."*
Bottom pair: N-39 Stainless knife and fork slide together in plastic handle for storage. Cap lifter on fork. Marked PAT.PEND MADE IN U.S.A. Both knife and fork marked STAINLESS. 7 1/2". Two examples shown. $15-25.

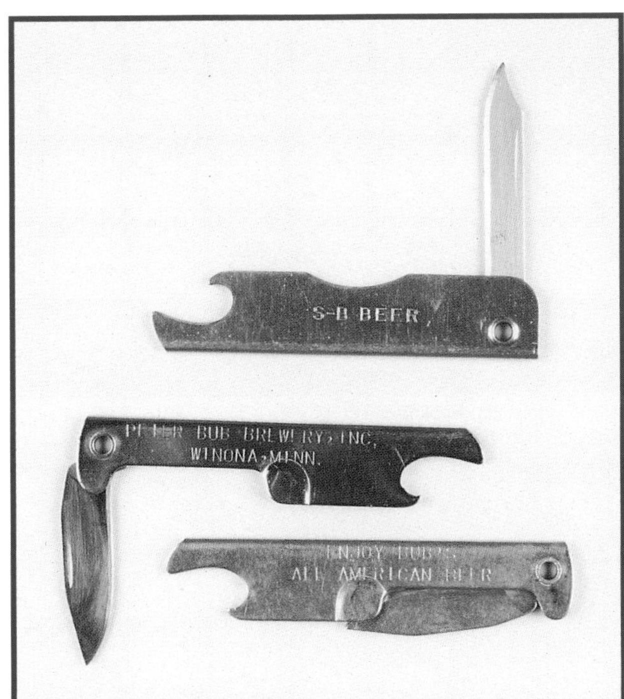

Top: N-35 Single blade knife with cap lifter formed in handle. Blade marked B&B ST. PAUL. 3". $25-30.
Bottom pair: N-43 Single blade knife with cap lifter formed in handle. 3 1/4". Shown open and closed. $20-30.

Top: N-44 Knife with master blade and cap lifter blade. Copper handles. Made by Dolphin Cutlery, New York. 3 1/2". $100-125.
Middle: N-61 Knife with master blade and cap lifter blade. Marked H. B. HARDENBERG & CO., GERMANY. 3 5/8". $75-90.
Bottom: N-87 Knife with master blade and cap lifter blade. Marked D PERES SOLINGEN GERMANY. 3 3/8". $75-90.

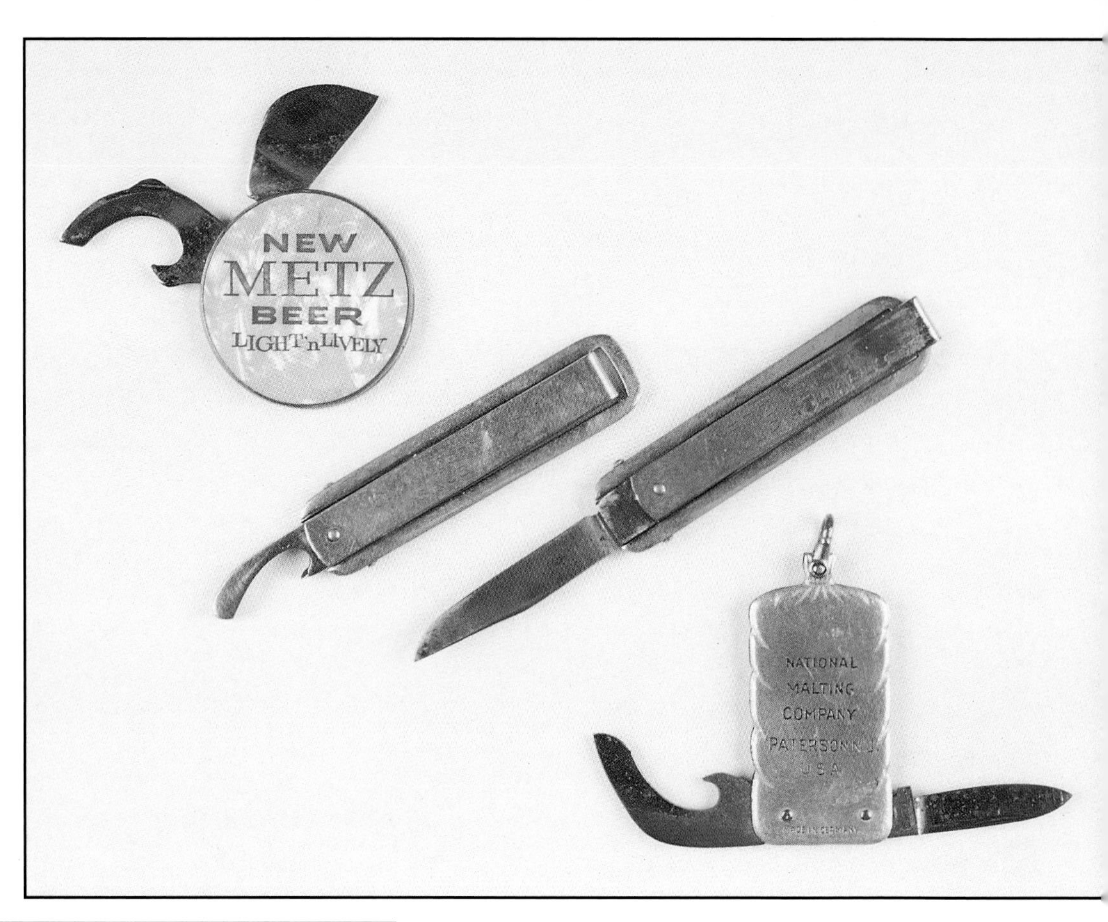

Top: N-45 Round knife with cutting blade and opener blade. American design patent 119,965 issued to Antonio Paolantonio, April 16, 1940. 1 1/2" diameter. $10-15. **Middle pair:** N-50 Single knife blade with cap lifter formed on back of blade. Blade swivels to expose opener or knife. 3 3/4" with opener exposed. $40-50. *A foreign example celebrates the fiftieth anniversary of San Miguel Brewery, 1890-1940.* **Bottom:** N-59 Sack of malt shape knife with pen blade and cap lifter blade. Made by Paul A. Henckels, Germany. 2 1/8". $40-50.

Top: N-47 Knife with master blade and cap lifter blade. Marked H. BOKER & CO., SOLINGEN. 3". $75-90. **Middle:** N-68 Knife with master blade and cap lifter blade. Marked SCHATT & MORGAN CUTLERY CO. TITUSVILLE, PA. 3 1/4". $75-90. **Bottom:** N-73 Knife with pen blade and master blade with cap lifter at the base. Marked CHRISTIANS SOLINGEN/RCM 3 5/8". $50-60.

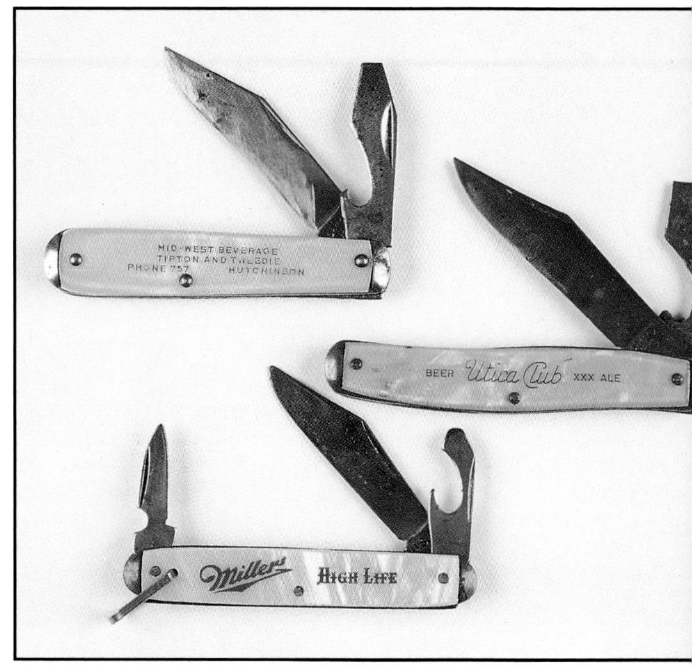

Top: N-53 Knife with master blade and cap lifter blade. Marked THE IDEAL, U. S. A. or CAMCO USA. 3 1/4". $35-40. **Middle:** N-60 Knife with master blade and cap lifter blade. Made by Utica Cutlery. 3 1/4". $35-40. **Bottom:** N-72 Knife with master blade, pen blade, and cap lifter blade. Marked AM. CO. U. S. A. 2 3/4". $35-40. *A souvenir piece for the "Falstaff Bowling League 1960" has been found.*

Top left: N-48 Bottle shape knife with master blade and cap lifter blade. Marked LUNAWERK, SOLINGEN, GERMANY. 3". $75-90. **Top right:** N-54 Bottle shape knife with master blade and cap lifter blade. Celluloid handles. Marked REMINGTON TRADE MARK and REMINGTON UMC MADE IN U. S. A. 3 1/4". $100-125. **Bottom left:** N-55 Bottle shape knife with master blade and cap lifter blade. Made by Paul A. Henckels, Solingen, Germany. 3 3/8". $100-125. **Bottom right:** N-80 Bottle shape cap lifter with knife blade. Marked OBERMEYER SOLINGEN GERMANY. 3 5/8". $75-100.

***Above:* Top:** N-63 Single blade knife with cap lifter on blade. 2 3/4". $200-250. *Advertises "Anheuser-Busch, Inc., Cabinets & Refrigerator bodies."* **Bottom:** N-70 Single blade knife with cap lifter formed in handle. Prest-O-Lite key. 3 1/4". $75-100.

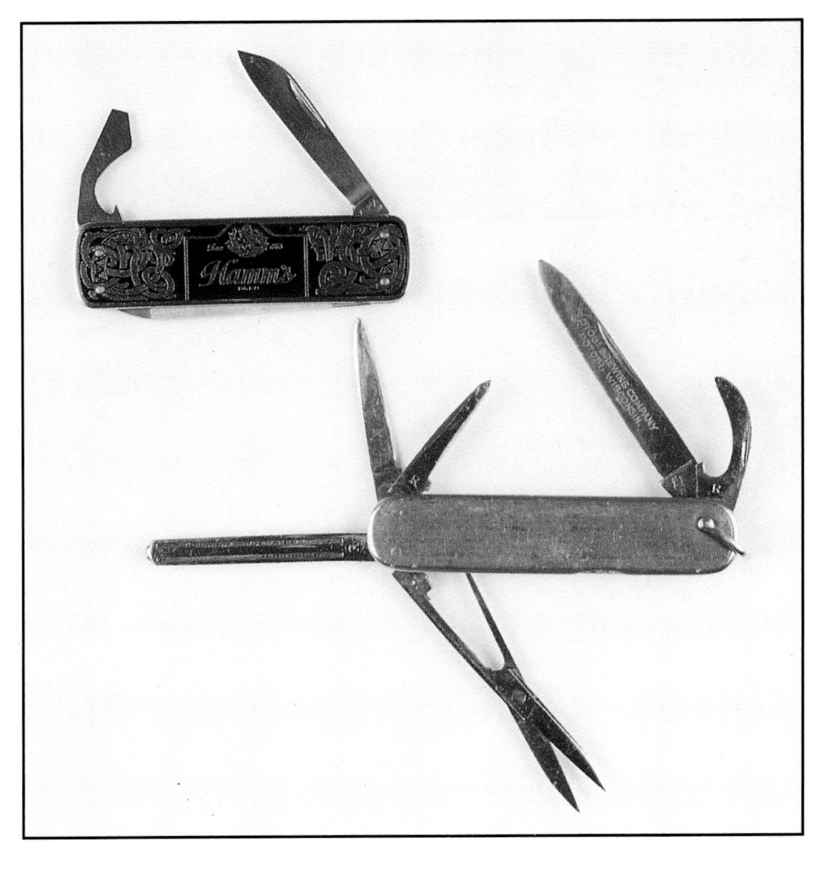

***Right:* Top:** N-74 Knife with master blade, file blade, and cap lifter blade. Ornate handles. Marked EKA, ESKILSTUNA. Eskilstuna is the cutlery center of Sweden. 2 1/8". $75-90. **Bottom:** N-92 Knife with cap lifter, three cutting blades, file blade, and scissors. Marked MADE BY BERTRAM, GERMANY ROSTFREI. 3 1/8". $75-100.

Top left: N-81 Triangular knife with cap lifter, pen blade, and key ring. Marked RUST FREE. 1 3/4". $5-8. **Bottom left:** N-89 Swiss Army knife with two opener blades and two cutting blades. Marked OFFICIER SUISSE. 3 1/4". $12-15. **Right:** N-90 Swiss Army knife including cap lifter, screwdriver, cutting blade, file blade, toothpick, and tweezers. 2 1/4". $12-15.

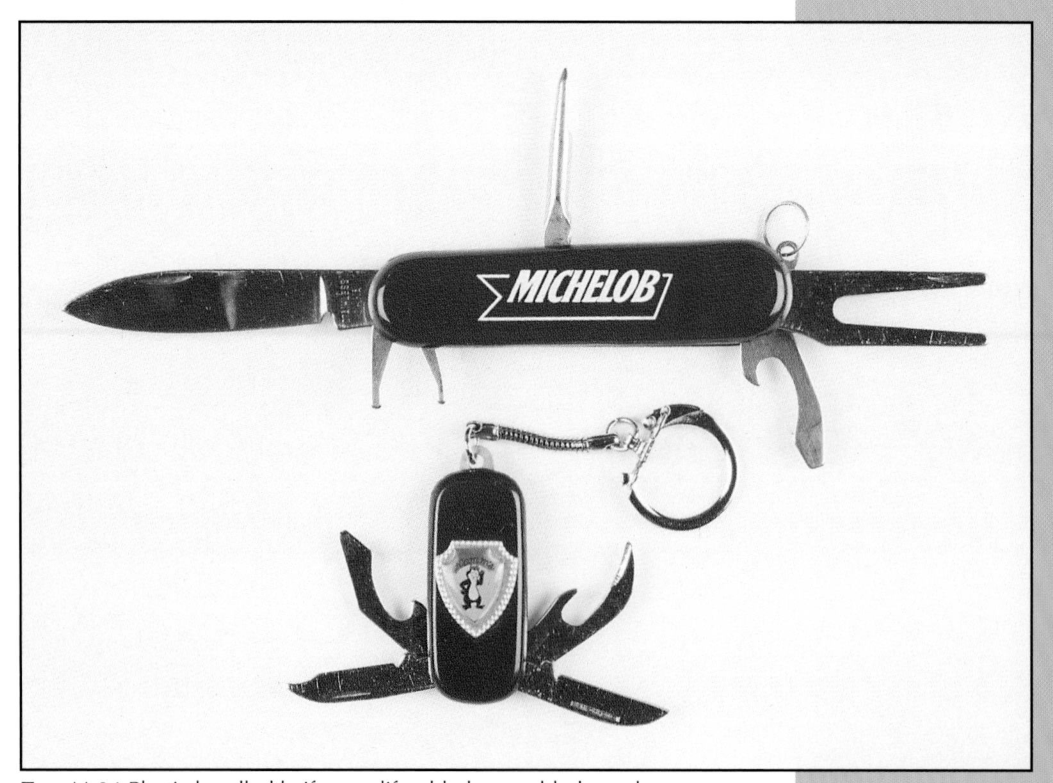

Top: N-94 Plastic handled knife, cap lifter blade, pen blade, awl, two spike wrench blades, and ball mark repair blade. 3 1/4". $8-10. **Bottom:** N-93 Black plastic handled knife, two cap lifter blades, pen blade, nail file blade, and key ring. Made in China. 2". $8-10.

Wall Mount Stationary Openers

At one time, the wall mounted bottle cap remover was a standard fixture in roadside motels. The owners screwed them to the wall in hopes that they would still be there at check out time—much less of a risk than leaving an opener on the dresser that could conveniently be pocketed on the way out. It was also insurance against the whims of the thirsty traveler who might otherwise resort to ripping a cap off using the underside of the bathroom counter or a hinge on a door. Although many wall mounts were produced with beer advertising, the traveler was more likely to find a plain example or one advertising Coca-Cola on his bathroom wall.

Cap lifters were also mounted in handy spots behind the tavern bar to save the bartender the constant frustration of locating an opener. This is where the wide variety of beer advertising wall mounted openers gained their fame.

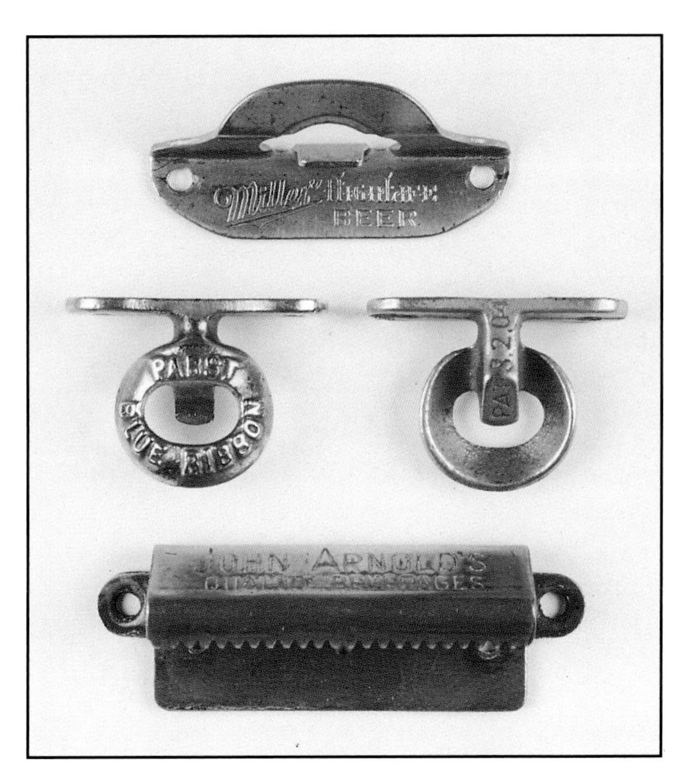

Top: O-1 Vaughan's "No-Chip" model #163 wall mounted cap lifter with two screws. 2 7/8" wide. $40-50. **Middle pair:** O-12 Wall mounted cap lifter mounts with two screws. Marked PAT 8,2,04. 2" wide. Front and back views shown. $100-125. **Bottom:** O-18 Toothed cap lifter mounts with two or three screws. Made by the Protector Mfg. Co. 4" wide. $100-125.

Left: O-2 Enameled wall mounted cap lifter with four screws. Made by Erickson Company, Des Moines, Iowa. 2 1/8" wide. $10-25. **Right:** O-6 Enameled wall mounted cap lifter mounts with three screws. Probably made by Erickson Company, Des Moines, Iowa. 2 5/8" wide. $30-50.

Left: O-3 The "Clipper" two piece plastic and metal wall mounted cap lifter with one screw. Marked REMEMBRANCE, B & B (Brown & Bigelow). 2 1/8" wide. $15-20. **Middle:** O-7 Metal and plastic wall mounted cap lifter. Bin catches caps. Made by B & B, St. Paul, Minnesota (Brown & Bigelow). 2 5/8" wide. $15-20. **Right:** O-14 Wall mounted cap lifter with bottle cap (to show what it's for!). 1 1/4" wide. $20-25.

Advertising for the O-3 Clipper Bottle Opener describes it as: "The Clipper Bottle Opener is made to do its job neatly and efficiently. It not only removes bottle caps in a jiffy, it hangs on to them until you want to slide them out through the opening on the side. Featuring nickel-plated, case-hardened steel grips, the unit is easily attached to the wall with one screw."

Top three: O-4 Vaughan's "Never Chip" models #1 and #2 wall mounted cap lifter. Mounts with two screws. American patent 1,029,645 issued to Harry L. Vaughan, June 18, 1912. The opener appears in Vaughan's 1922 and 1970 catalogs. Single packed in box claiming "It's the only *Stationary* Bottle Opener made which will remove 'the cap' without chipping the bottle, it's flexible." 2 1/2" wide. $15-30. *Do you know how to find a good beer? Try this message: "Enjoy your Gluek's. The beer for the man who knows."* **Bottom pair:** O-5 "Starr" wall mounted cap lifter with two screws. Manufactured by Brown Manufacturing Co., Inc., Newport News, Virginia. Trademarked as STARR X. American patent 2,033,088 issued to Raymond M. Brown, November 2, 1943. In a 1946 advertisement, we learn that the Starr is "The World's Best Opener. Eliminate loss of bottles and contents. Prevent danger to the public. Have long life." 2 3/4" wide. Two examples shown. $5-75.

Left: O-8 Wall mounted cap lifter with four screws. American patent 1,711,678 issued to Thomas Harding of Newark, New Jersey, May 7, 1929. Produced by J. L. Sommer Manufacturing Company. Also covered by Canadian patent 289,495. 2" wide. $30-40.
Middle: O-10 Wall mounted cap lifter mounts with three screws. 2" wide. $50-60.
Right: O-19 Wall mounted cap lifter mounts with three screws. 3 3/8". $30-40.

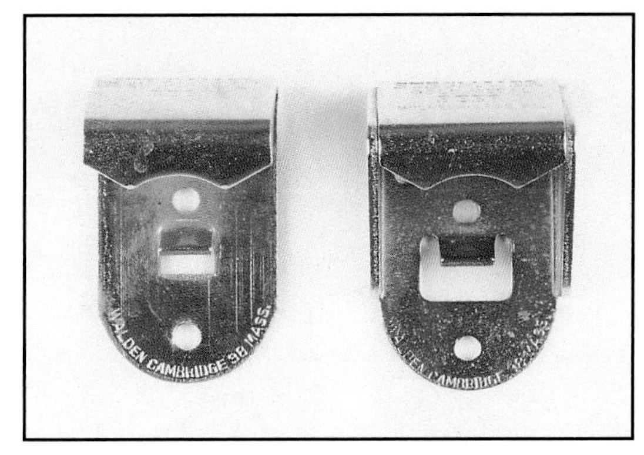

Left: O-9 Wall mounted cap lifter mounts with two screws. No side walls. Marked WALDEN CAMBRIDGE 38 MASS. American design patent 160,453 issued to Davis J. Ajouelo, October 17, 1950. Depicted in a 1961 Handy-Walden, Inc. (New York) catalog and called "Wal-Ope." 1 1/4" wide. $15-20.
Right: O-16 Wall mounted cap lifter mounts with two screws. Side walls. Marked WALDEN CAMBRIDGE 38 MASS. 1 3/8" wide. $15-20.

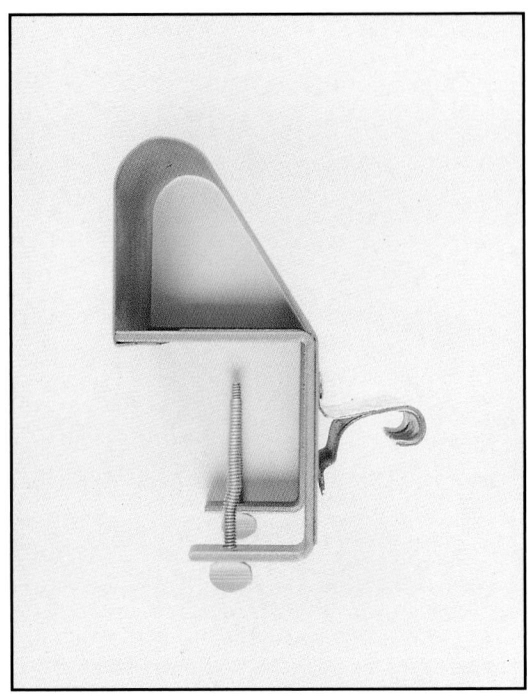

O-11 Bar mounted opener with two thumb screws. Has Vaughan's "Never-Chip" (O-4) opener attached to large bracket plate. 5" wide. $60-75.

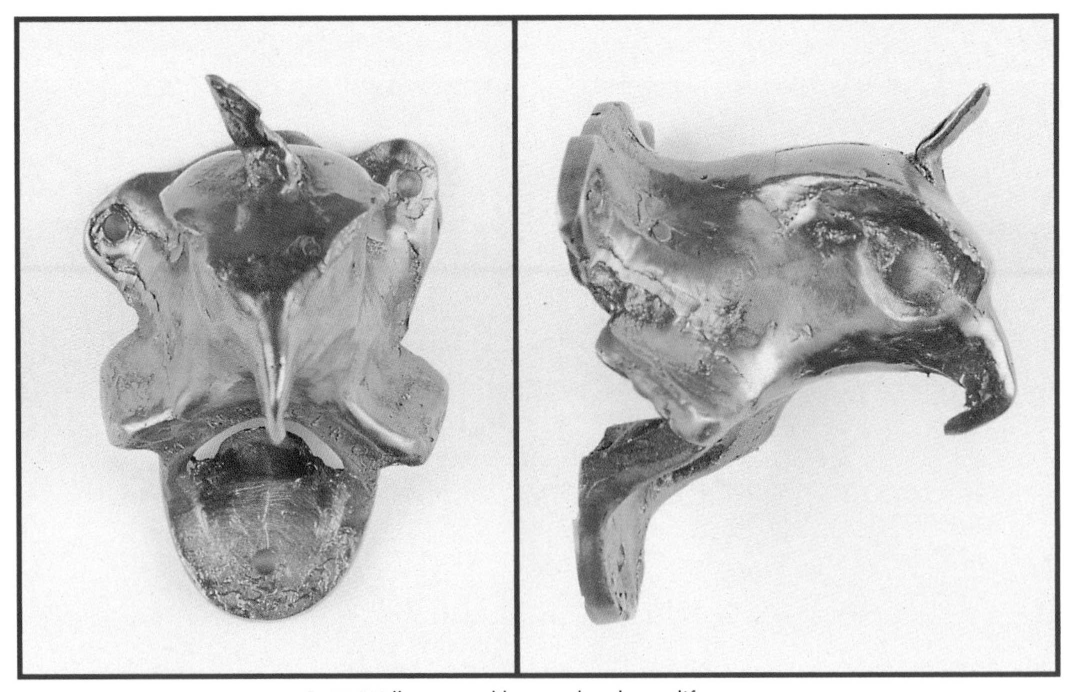

O-13 Wall mounted bronze hawk cap lifter. Marked MENDOCINO. 2 3/4" wide. $20-25.

O-15 Wood plaque with barrel on ledge. Cap lifter is under ledge. 4 1/2" tall and 3 1/4" wide. $40-50.

O-17 Type O-5 cap lifter mounted inside metal crown cap catcher. 9" tall and 5 1/2" wide. $20-25.

Corkscrews

Corkscrews played an important role in beer advertising and as a necessary tool in the 1800s and into the early 1900s. Prior to the invention of the crimped bottle cap, breweries used corks that were secured by wire or string to the bottle. The corks protected the precious brew inside. For many years, the corkscrew was the opener for beer as well as medicines, soda waters, inks, photographic liquids, perfumes, whiskey, and wine. The bottle cap did not become commercially available until the late 1890s and many skeptical brewers continued to favor the cork until the early 20th century.

When searching for beer advertising corkscrews, the advertising collector will often butt heads with the corkscrew addict. These addicts are usually relentless in pursuit of corkscrews types that will fill holes in their collections. To them it doesn't matter whether there is beer advertising or not. Meanwhile, the beer advertising corkscrew collector may desperately want to add another brand name to a type he already owns. The struggle begins when they both need the same corkscrew. What they will ultimately find is that some corkscrews will bring a higher price amongst corkscrew collectors for the type, while advertising collectors may be willing to pay a higher price for the brand name on any example. This dilemma is reflected in the wide range of values shown on types in this category.

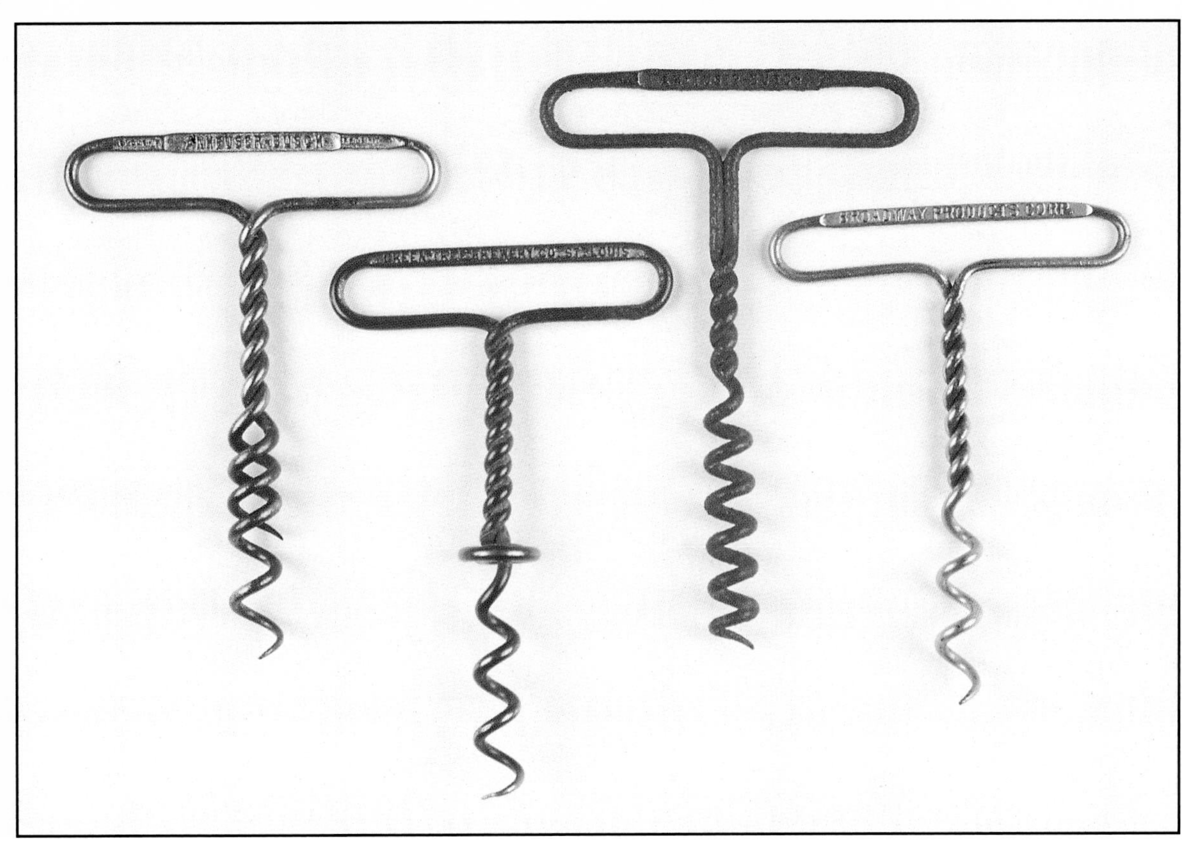

Left to right: P-1 "Duplex Power Cork Screw." American patent 172,868 issued to William R. Clough, February 1, 1876. $40-100. P-73 Steel wire corkscrew. Double and twisted wire shank. Stopper button. American patent 172,868 issued to William R. Clough, February 1, 1876. $75-125. P-88 Steel wire formed corkscrew with twisted shank and single helix. $75-125. P-94 Wire handle with twisted shank and single helix. $75-100.

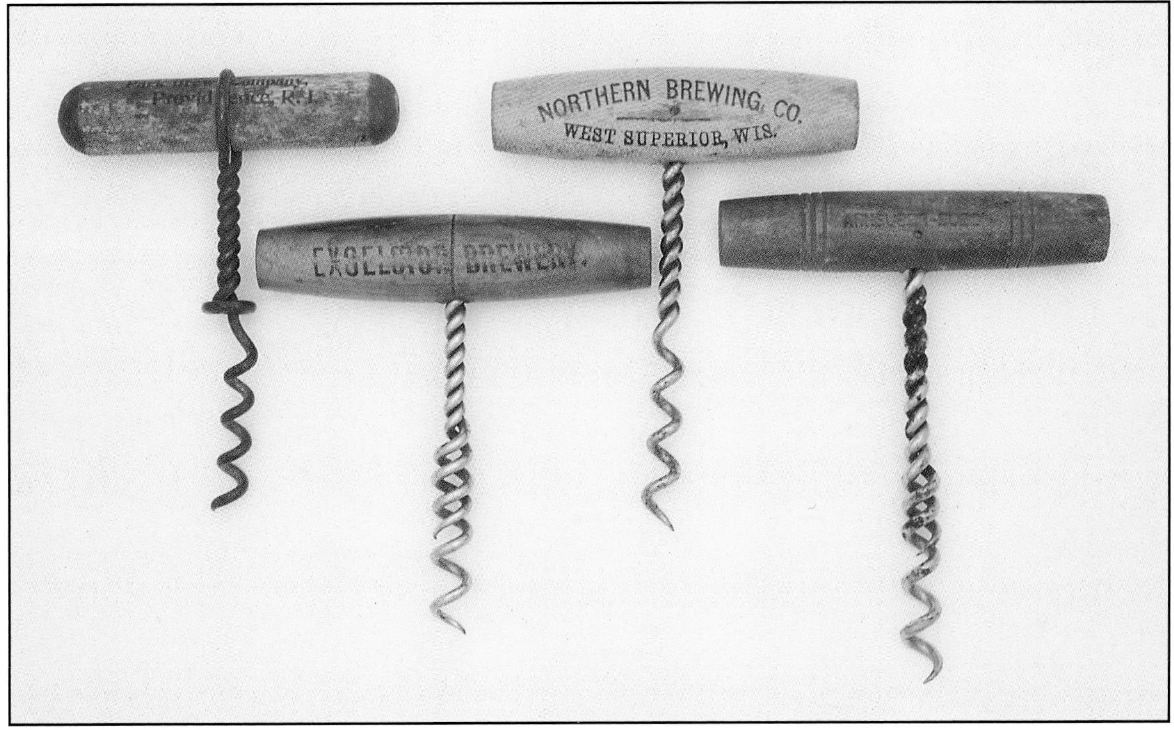

Left to right: P-60 Twisted wire corkscrew, single helix. Stopper button. Wire loops around handle. $40-50. P-61 Twisted wire corkscrew, double helix. Handle pinned to top of shank. $30-50. P-90 Wood handle twisted wire corkscrew with single helix. $50-60. P-127 Twisted wire double helix corkscrew. Handle has six concentric rings. $100-125.

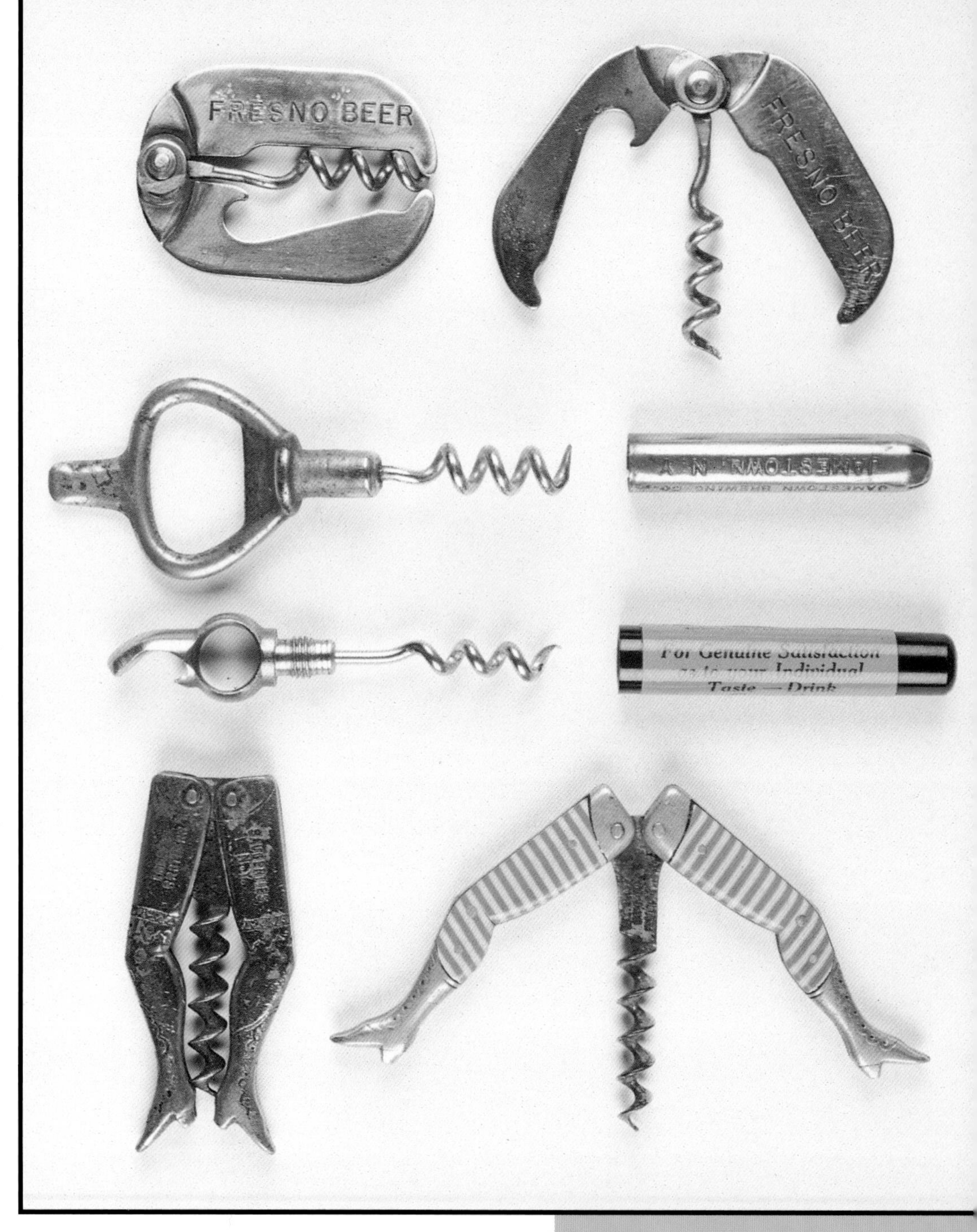

Top pair: P-11 The "Tip Top" by Williamson of Newark, New Jersey. A packaging card for the Tip Top proclaims "Lies flat in your vest pocket." Shown opened and closed. $125-150. **First middle:** P-18 Cap lifter with screwdriver tip and cork-screw in metal sleeve. $50-75. **Second middle:** P-63 Picnic corkscrew with cap lifter. Marked NO. 19Y CORK SCREW & BOTTLE OPENER. $100-125. **Bottom left:** P-79 Folding gay nineties legs. Nickel plated brass. Made in Germany. $400-500. **Bottom right:** P-114 Folding gay nineties legs. Celluloid, striped. German patent 21718 issued to Steinfeld & Reimer, January 1, 1894. A 1910 Norvell-Shapleigh Hardware Company catalog calls the legs "Ballet" and offers them at $14 per dozen. $400-500.

P-3 Collapsing corkscrew. American patent 447,185 issued to Carl Hollweg, February 24, 1891. Hinges in four places allow the user to fully close the corkscrew for storage or fully open to use the case as the handle. A 1913 catalog from Lewis Brothers of Montreal, a hardware wholesaler, offered this corkscrew as the "Telescope." They sold them for $2.70 per dozen. $75-100.

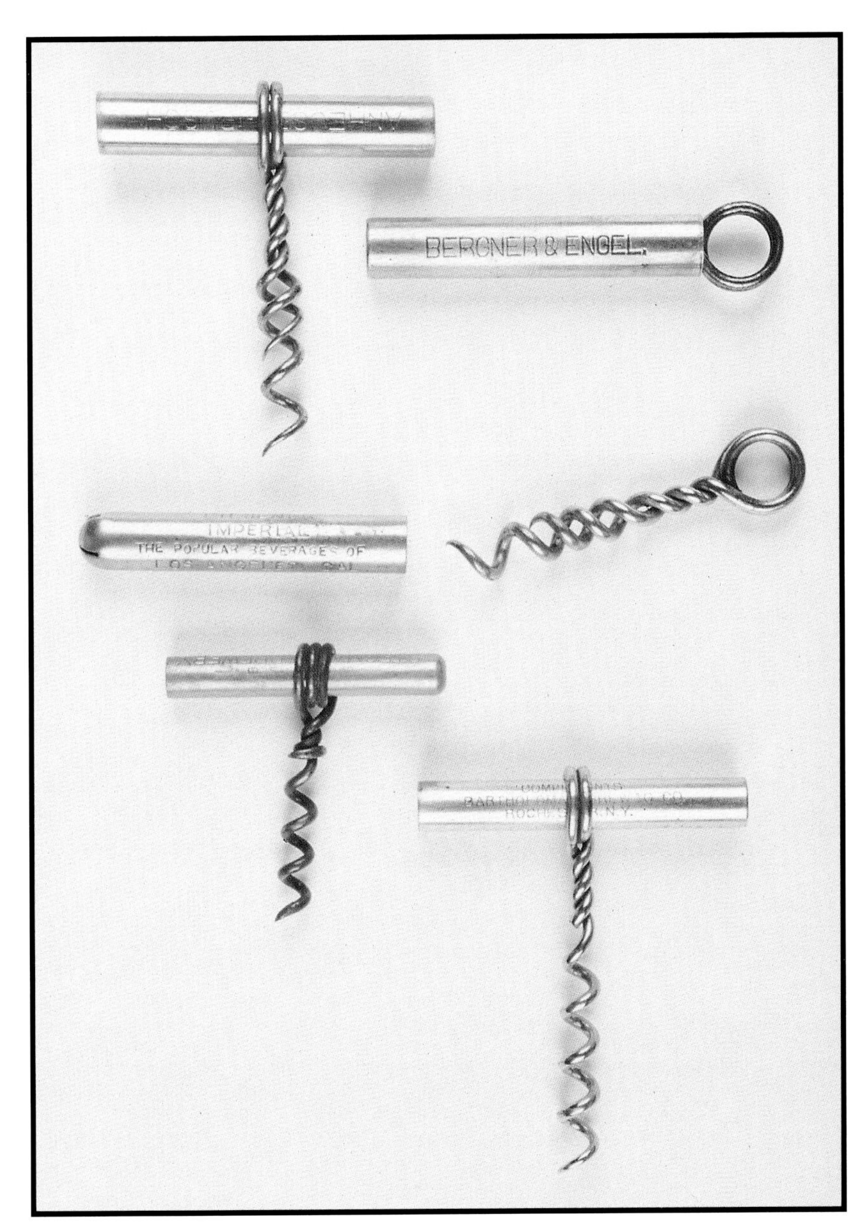

***Left:* Top pair:** P-5 Picnic type corkscrew. Double twisted wire screw. Made by C. T. Williamson Wire Novelty Company, Newark, New Jersey. Called "Williamson's Power Pocket Corkscrew." Some marked WILLIAMSON. Some have left-hand worms. Sheath is 2 3/4" x 7/16". $50-125. **Middle:** P-103 Picnic corkscrew. Closed end metal sleeve protects worm. Made by C. T. Williamson Wire Novelty Company, Newark, New Jersey. Sheath is 2 3/4" x 7/16". $50-100. **Bottom left:** P-62 Picnic corkscrew with nickel plated brass sheath. Double twisted wire worm. Closed end barrel. Sheath is 2 1/4" x 5/16". Two views shown. $75-100. **Bottom right:** P-140 Picnic type corkscrew. Sheath is 2 11/16" x 3/8". $50-100.

***Below:* Left to right:** P-4 Cap lifter/corkscrew with bell. Wooden sleeve protects worm. $75-100. P-6 Cap lifter/corkscrew with wooden sheath. $25-50. P-31 Cap lifter with wire breaker and corkscrew. Wooden sheath protects worm. $15-30. P-53 Corkscrew with two fret two piece wire formed cap lifter. Wood sheath protects worm. $10-30. P-74 Two fret one piece wire cap lifter and corkscrew. Worm protected by wooden sheath. $20-40. P-85 Cap lifter with screwdriver tip and corkscrew. Worm protected by wooden sheath. $20-40.

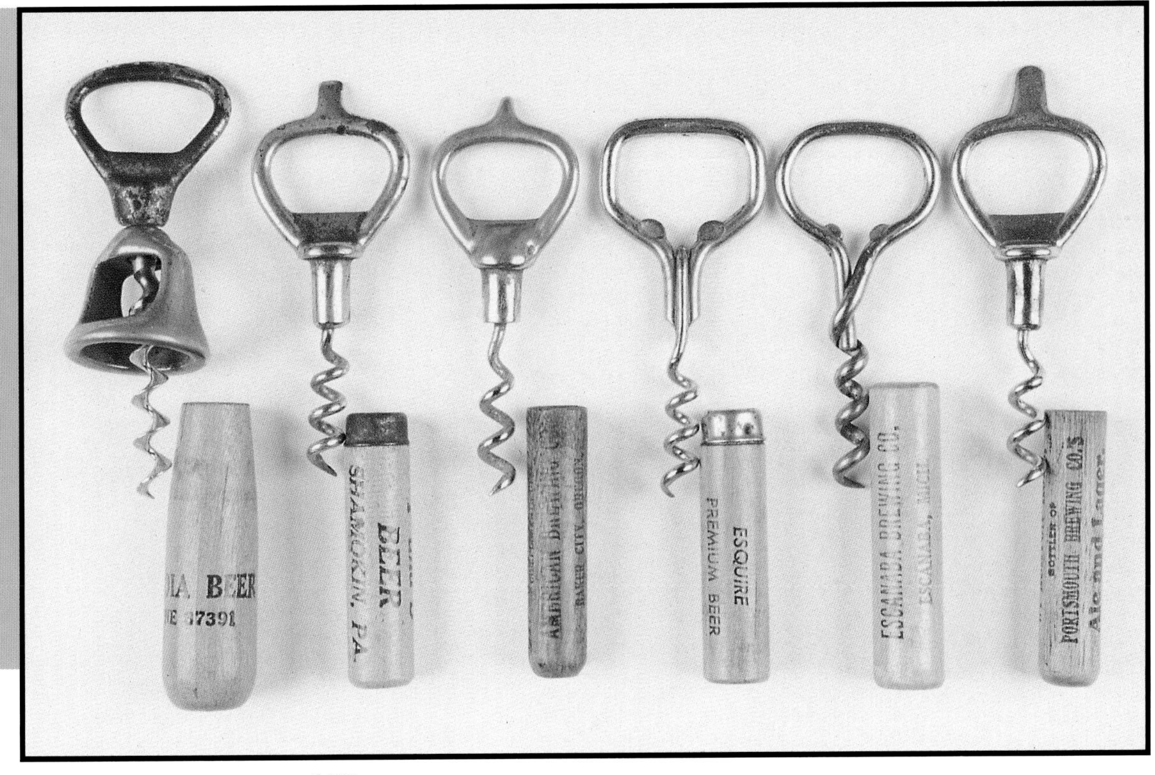

Top: **Left side:** P-19 Single/double ring wire handle corkscrew with wooden sheath (made by the ga-zillions). $10-30. **Right side:** P-54 Single/double ring wire handle corkscrew with "Decapitator." $15-50. *An interesting advertisement on this type was: "The one thing finished in this nasty world is P. B. Ale."*

Center: P-7 The "All-Ways" handy combination bottle opener and corkscrew. Advertised as "No two ways about this being useful, it's useful in four ways - Pulling a Cork, Taking out Aluminum Stopper, Removing a Seal, Lifting a Crown Cap." 1900 patent date on these is a reference to Clough's machine for bending wire into a corkscrew. A 1901 patent date on these is John Baseler's patent for a "Stopper Extractor." The patent was for the cap lifter with the point at the front end of the crescent designed to punch a hole in the cap so it could not be re-used. A 1916 *Western Brewer* magazine advertisement from A. W. Stephens Company of Waltham, Massachusetts, proclaims "one of these openers hung up in the kitchen beats a hundred of the other kind scattered on the cellar floor." Supplier names are usually included in small print and often with the incorrect patent dates March 30, 1901 and April 30, 1910. One advertiser took full advantage of the space allowed by using 64 words to say "Sole distributors in Boston for Piel Beer. Draught and brewery bottling. Headquarters for Holland Gin. Importers of Ports, Sherries, Brandies, Irish, Scotches, and Gins. Holland System, Inc., 47 Boylston St., Boston, Mass. Sole owners Green River Rye. Sole owners system beer and ale. Brewery bottling, every bottle a new bottle. Sole agents U. S. A. Shamrock Stout from Dublin Ireland. Draught and brewery bottling." $25-50.

Bottom: **Top to bottom: (pair)** P-33 Cap lifter with wire breaker and corkscrew. Wooden sheath to protect worm is also handle of ice pick. Two views shown. $25-50. P-138 Wire two-fret cap lifter and corkscrew. Wooden sheath to protect worm is also handle of ice pick. $50-60. P-133 Double ring corkscrew inserts in end of ice-pick handle. Cap lifter between ice-pick and handle. Marked PAT. PEND. Supplied by IRA F. WHITE & SON CO. BLOOMFIELD, N. J. $60-75. P-68 Can opener with sliding adjustable blade and Clough wire corkscrew in the handle. Marked "SURE-CUT" CAN OPENER PAT 07-19-04. American patent 765,450 issued to Frank White and Fred Winkler, July 19, 1904. $50-75.

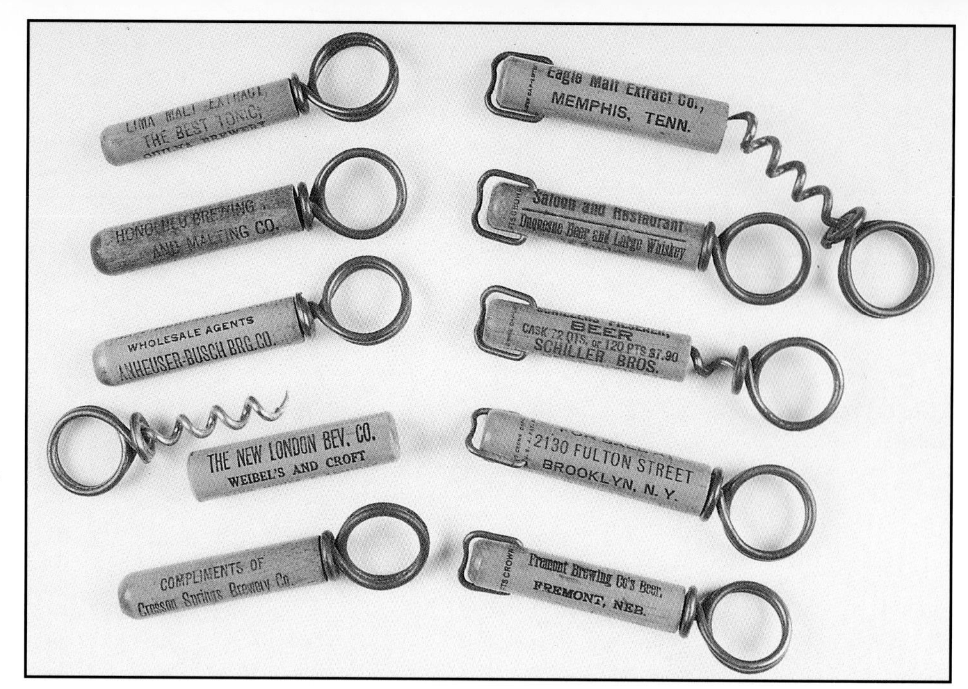

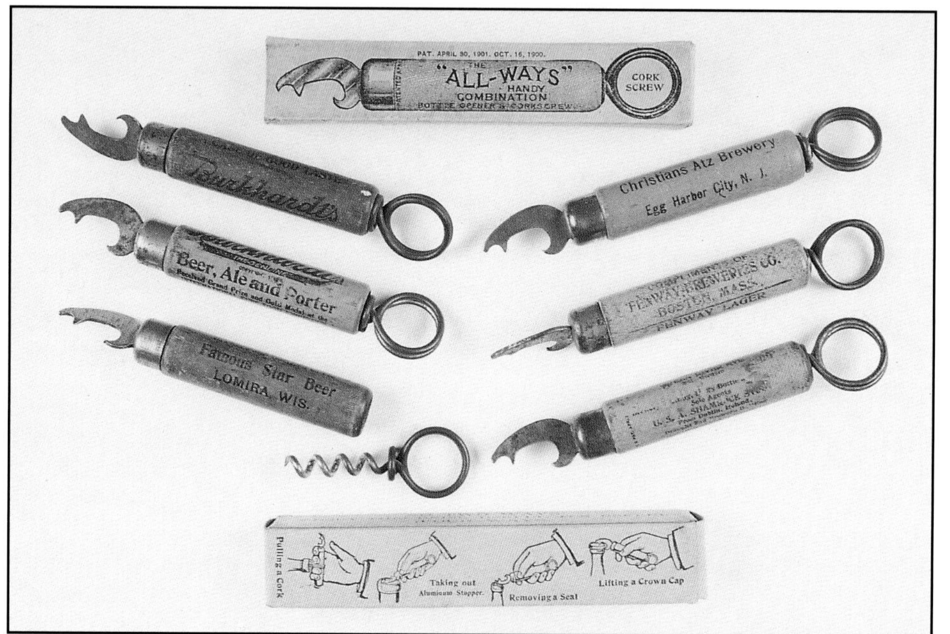

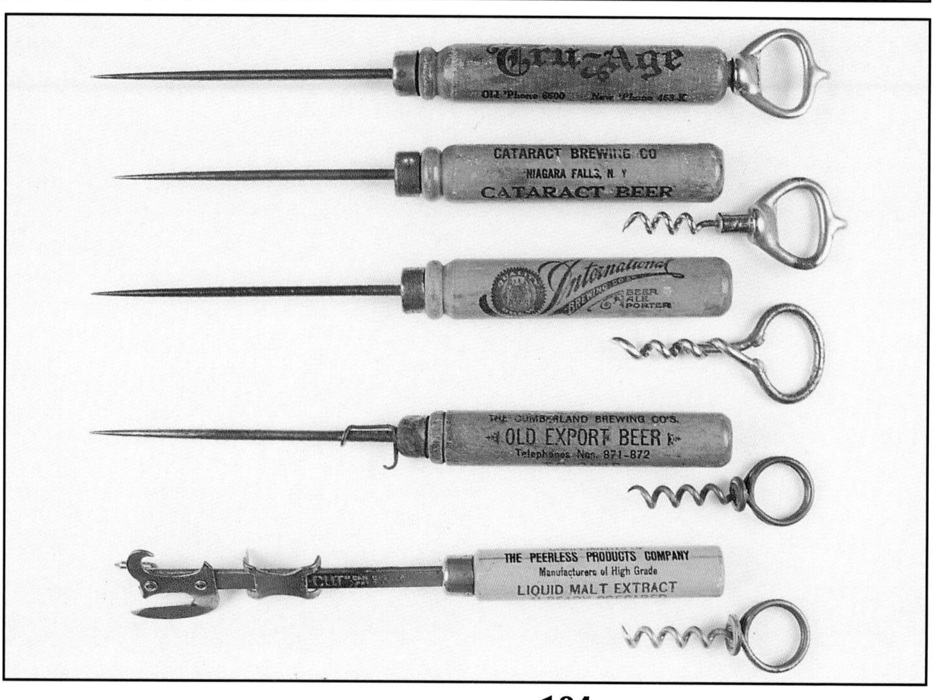

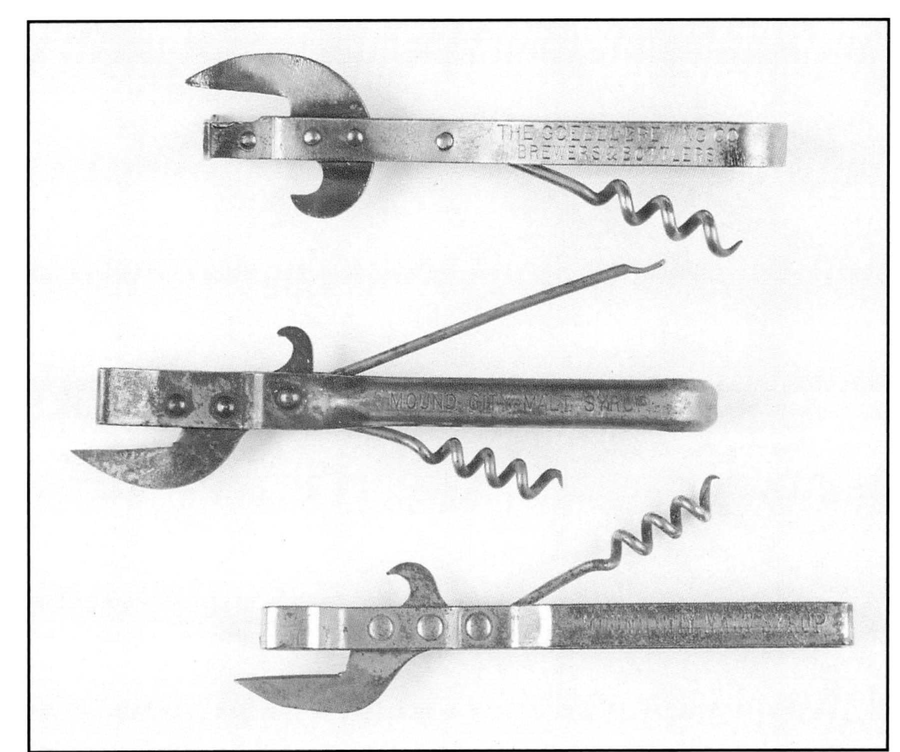

Top: P-23 "Yankee" can opener with cap lifter and folding corkscrew. Rounded handle end. American patent 839,229 issued to Charles G. Taylor, December 25, 1906. $30-60. **Middle:** P-37 Can opener with cap lifter, corkscrew, and cocktail fork. $75-100. **Bottom:** P-106 Can opener with cap lifter and corkscrew. Squared end handle. $35-50.

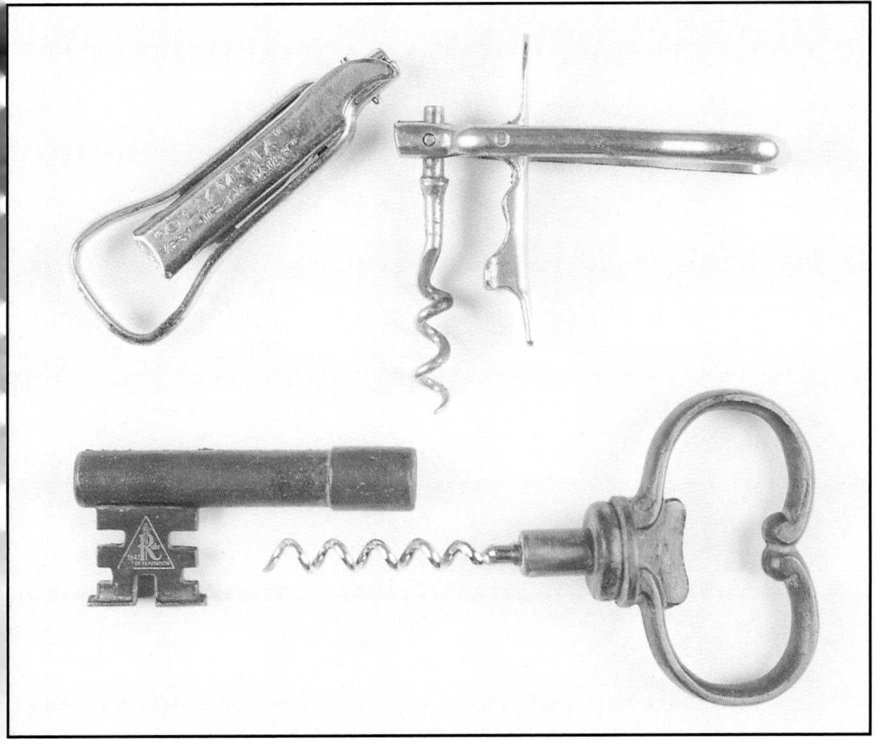

***Above:* Top:** P-41 "Universal" folding corkscrew and cap lifter. American patents 793,318 and 824,807 issued to Harry W. Noyes, June 27, 1905 and July 3, 1906. Shown folded and ready for action. $50-75. **Bottom:** P-58 Key shape cap lifter with concealed corkscrew. $50-60.

Right: P-72 Ornate sterling corkscrew. Magnificently decorated example of an 1886 patent by Edward Thiéry and Charles Croselmire. The silver has been applied to an ivory handle with ornate floral design caps added to the end. There is fine detailing on all sections wrapping the handle. The silver is embossed with "Compliments Rochester Brewery" and a hand holding a glass of beer. It is from Rochester Brewery in Kansas City, Missouri. Marked STERLING 5827 without a maker's mark. c.1900. $600-800.

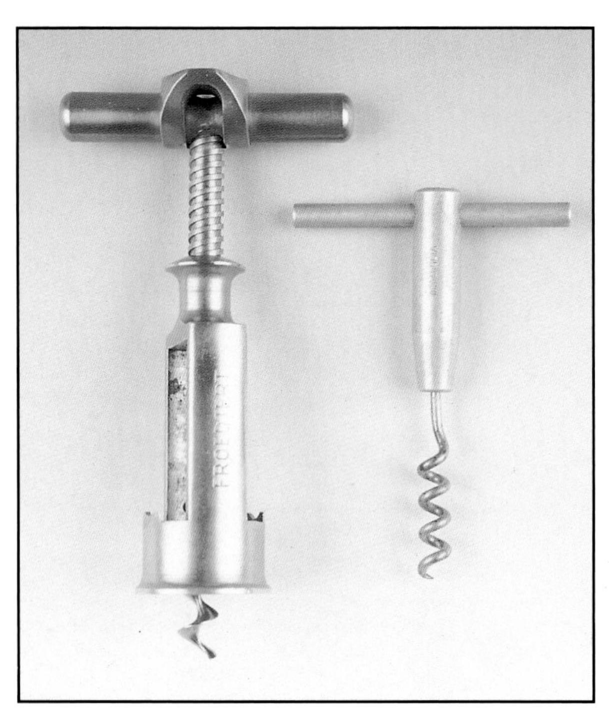

Left: P-80 Open frame corkscrew with locking handle with two cap lifters. $150-200. **Right:** P-126 T-handle metal corkscrew. $50-60.

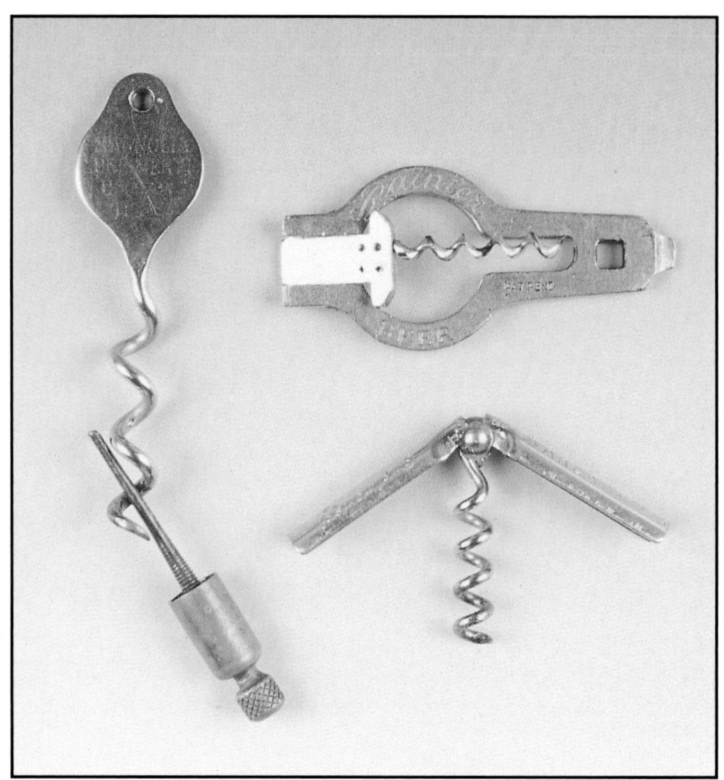

Left: P-87 Peg and worm corkscrew. The peg is stored in the center of the worm and removed and inserted into a hole at the top of the worm for use. In this type, the peg threads into the hole. American patent 611,046 issued to Edwin Walker, September 20, 1898. $150-250. **Top right:** P-100 Folding pocket corkscrew with cap lifter, screwdriver tip, and Prest-O-Lite key. Marked PAT. PEND. American patent 1,116,509 entitled "Cap Remover" issued to Josephine Spielbauer, November 10, 1914. $250-300. **Bottom right:** P-109 "The Dainty" folding corkscrew. Made by Vaughan Novelty Manufacturing Company, Chicago. $100-125.

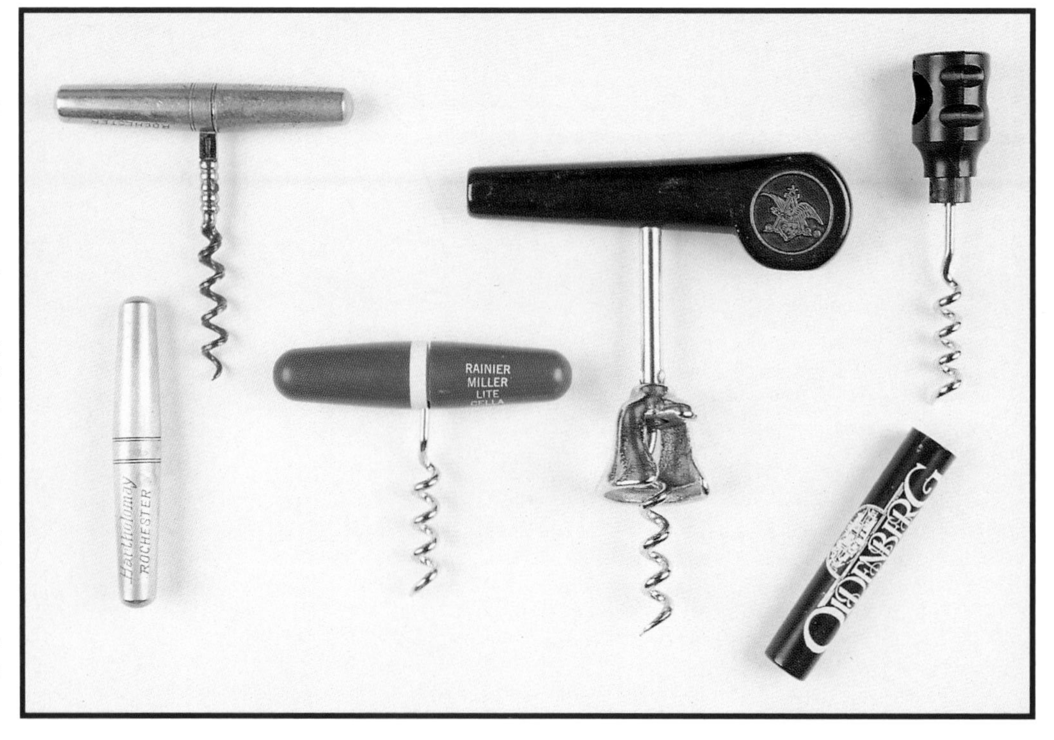

Left pair: P-44 Roundlet or Beau Brummel type pocket corkscrew. The case unscrews to reveal a worm stored in one end. The worm is pulled out and turned perpendicular to the case; then the empty side is threaded back on to form the handle. Shown open and closed. $100-150. **Left middle:** P-119 Roundlet or Beau Brummel pocket corkscrew. Plastic case. $10-15. **Right middle:** P-131 Plastic handle cork-screw. Wire breaker and cap lifter are part of cast bell. Anheuser-Busch Eagle-In-A medallion in handle. Part of a set-see M-85. Made by Williamson. $35-50. **Right:** P-98 Picnic corkscrew. Plastic. $3-5.

Left to right: P-8 Wood handle corkscrew with cast bell. American patent 501,975 issued to Edwin Walker, July 25, 1893. $15-50. P-57 Wood handle corkscrew with wire breaker on cast bell. $15-50. P-10 Wood handle corkscrew. Wire breaker and cap lifter are part of cast bell. American patent 647,775 issued to Edwin Walker, April 17, 1900. $15-50. P-118 Wood handle corkscrew with cast iron wire breaker above cast bell. $40-50. P-51 Wood handle corkscrew with cast iron wire breaker and cap lifter above cast bell. American design patent 29,798 issued to William A. Williamson, December 13, 1898. $15-50. *Charles Kaier Company of Mahanoy City, Pennsylvania put a new twist on corkscrews with this copy: "This is loaned. Must be returned."*

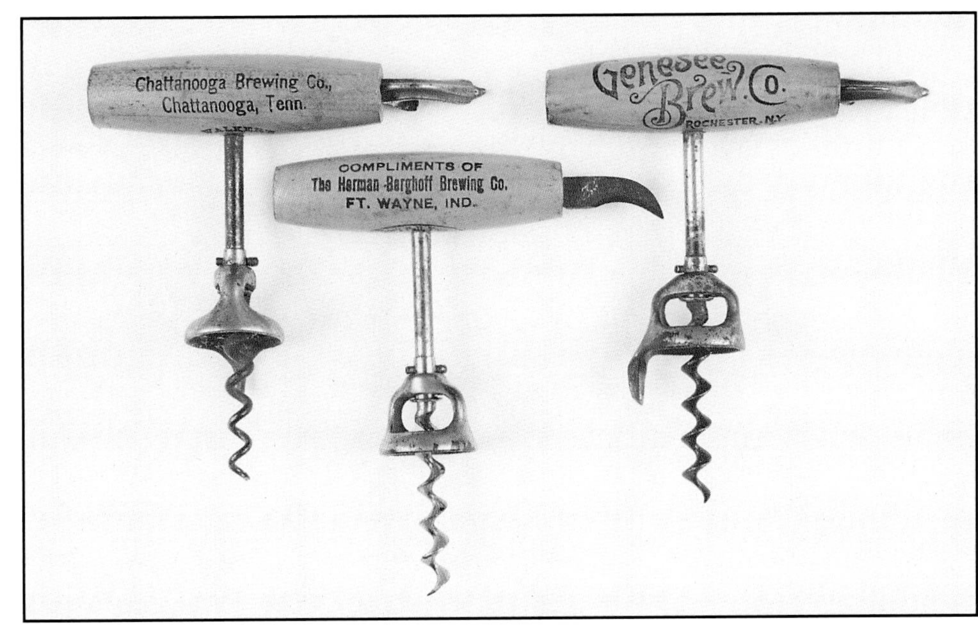

Left: P-9 Wood handle corkscrew. Cast iron crown cap lifter in handle. Walker 1893 patent. $25-60. **Middle:** P-17 Wood handle corkscrew with wire breaker/foil cutter in end of handle. Walker bell. American patent 579,200 issued to Edwin Walker, March 23, 1897. $30-60. **Right:** P-139 Wood handle corkscrew with wire breaker on cast bell. Cast iron crown cap lifter in handle. Walker 1893 patent. $40-50.

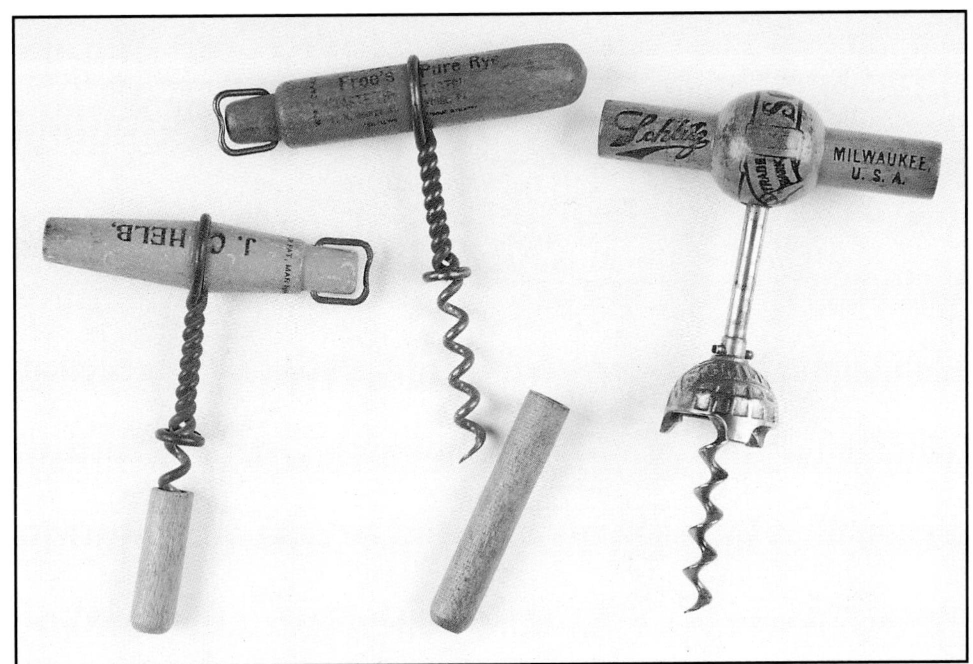

Left: P-12 Wood handle corkscrew with "Decapitator." Wood sheath protects worm. American patent 950,509 issued for the "Decapitator" to William R. Clough, March 1, 1910. $25-50. **Middle:** P-124 Wood handle corkscrew with "Decapitator." $60-75. **Right:** P-13 Schlitz self-puller corkscrew made by Erie Specialty Manufacturing Co. for Schlitz between 1888 and 1891. The globe in the center of the handle represents the Schlitz trademark used extensively in their advertising. It says "Schlitz Trade Mark." The sides of the handle read "Schlitz, Milwaukee, U. S. A." The word "Schlitz" is cast into the half globe bell. The end of the handle is marked E.S.M. CO., ERIE, PA., WALKER PAT. APPLIED FOR. This was the forerunner to Walker's 1893 patent. $150-250.

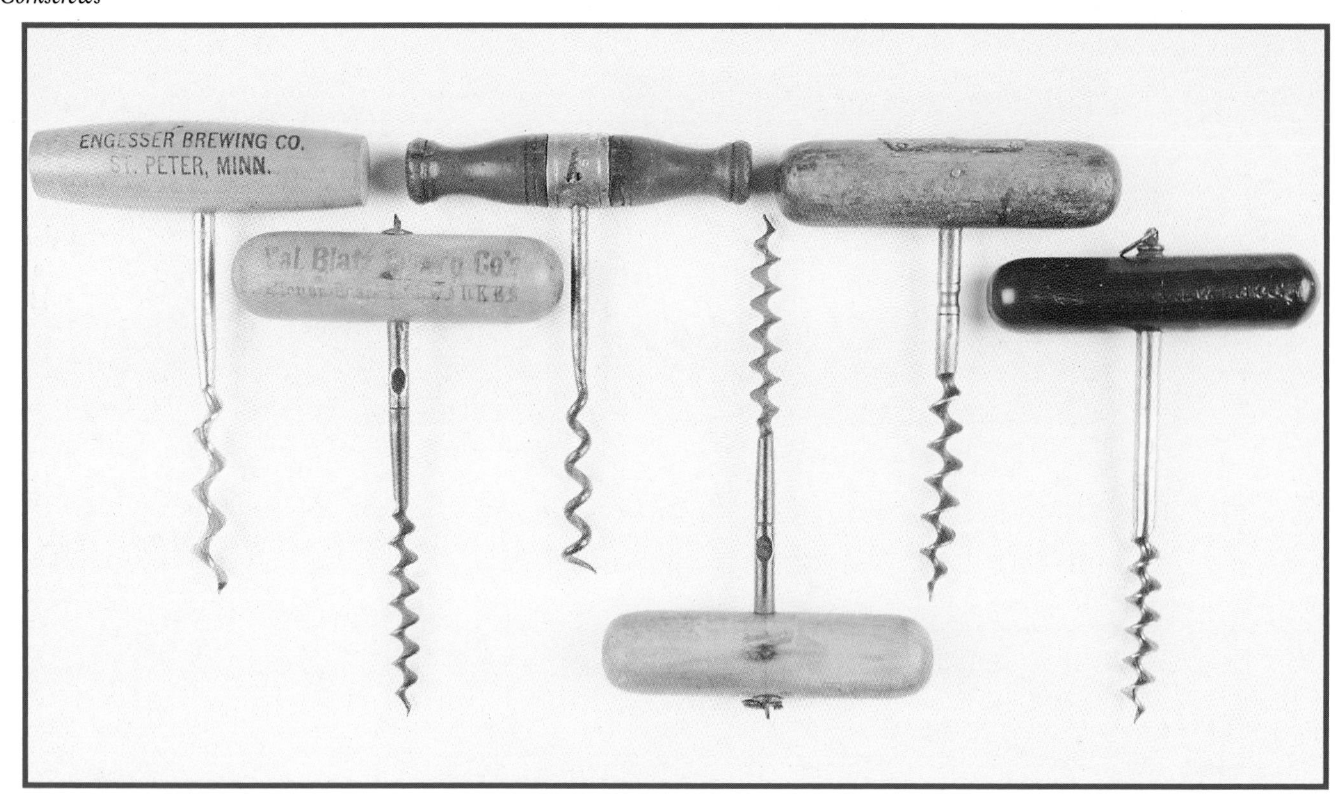

***Above:* Left to right:** P-22 Wood handle direct pull corkscrew. Made by Williamson. Some found with "speed worm." $15-30. P-50 Wood hot dog handle direct pull corkscrew. Made in Germany. $75-90. P-56 Wood handle corkscrew with brass band around handle. American patents 315,773 and 317,123 issued to Edward P. Haff, April 14 and May 4, 1885. $50-75. P-59 Barrel shape wood handle direct pull corkscrew. $60-75. P-77 Wood hot dog handle corkscrew with advertising plate affixed to the top. $75-100. P-78 Wood hot dog handle corkscrew with hook ring at top of handle. $75-100.

***Above:* Left:** P-24 Corkscrew with flat wood handle. Cast bell. Made by Williamson. $25-50. *One example is stamped "Drink Augusta Brewing Co's Export Beer" on one side and "Stroh's Beer, Detroit" on the reverse. This was most likely a production error. The same corkscrew is found with the Augusta Brewing message on both sides and others are found with the Stroh's message only.* **Middle:** P-43 Direct pull wood handle corkscrew, flat on both sides. Marked WILLIAMSON'S on the shank. $20-50. **Right:** P-132 Flat wood handle corkscrew with cast iron wire breaker and cap lifter above bell. Made by Williamson. $40-50.

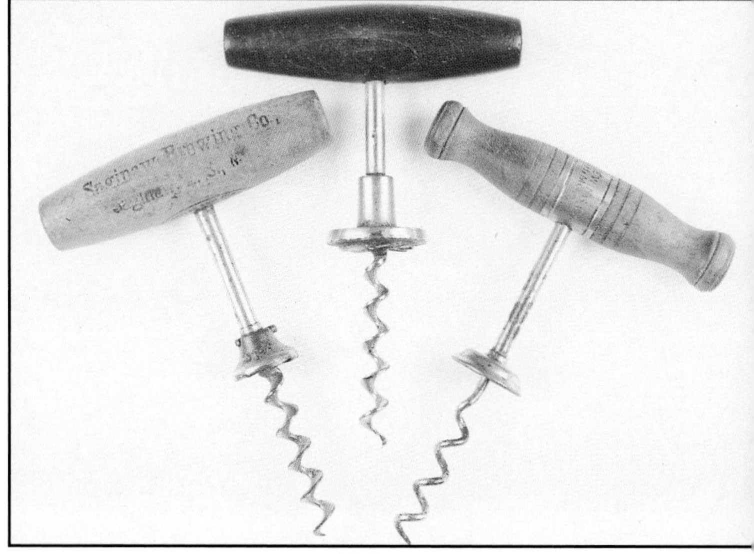

***Right:* Left:** P-32 Wood handle corkscrew with button. $150-175. **Middle:** P-76 Wood handle corkscrew with button. Advertising is cast into the top of the button. Made by Robert Murphy of Boston. $125-150. **Right:** P-86 Wood handle corkscrew with button on shaft. Ring around handle holds shaft. $75-100.

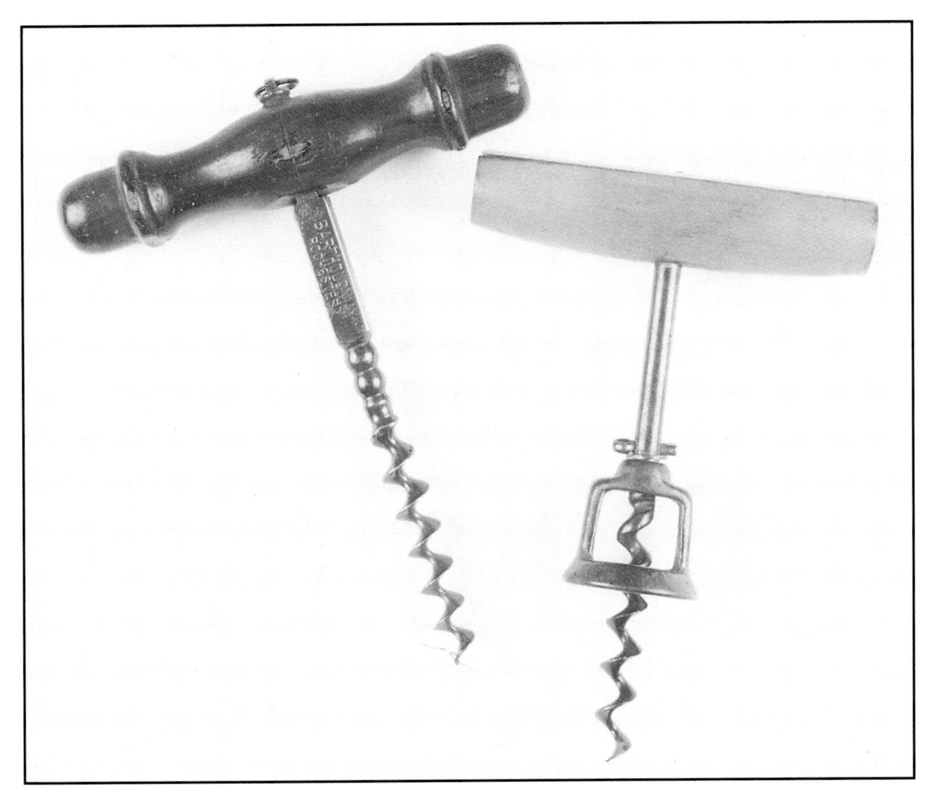

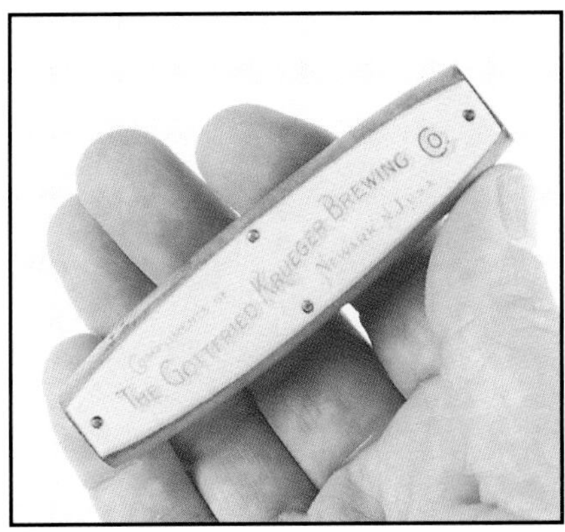

Left: P-42 Acorn handle corkscrew with square shank. $50-75. **Right:** P-67 Wood handle corkscrew with cast bell. The top of the handle is flat with a celluloid plate affixed with four small nails (*right*). $200-250. *The plate reads "Compliments of The Gottfried Krueger Brewing Co., Newark, N. J., U. S. A." Krueger was founded in Newark in 1852 by Braun & Laible. The company went out of business in 1961 and is probably best remembered as the company that introduced canned beer in 1935.*

Left to right: P-52 Wood handle corkscrew with spring barrel. Clip stops spring. Marked GERMANY. $100-125. P-115 Wood handle spring barrel corkscrew. Cotter pin on shank. Marked GERMANY. $75-100. P-141 Wood handle spring barrel corkscrew with paper clip type retainer above barrel. $150-200. P-75 Wood handle corkscrew with cast bell. Spring to retain cast bell instead of usual cotter pin. $100-125. P-93 Open frame corkscrew with spring between barrel and handle. $150-200.

Left: P-55 Wood handle corkscrew with heavy cast bell. $25-35. **Middle:** P-116 Wood handle corkscrew with cast iron wire breaker and cap lifter. Smaller version of normal wood handle corkscrews with bell - 3" wide and 4 1/4" long. Marked WILLIAMSON'S on shank. Marked WILLIAMSON CO. NEWARK, NJ, PATENTED DEC. 13 '98 on end of wood handle. American design patent D-29,798 entitled "Cap Lift" issued to William A. Williamson, December 13, 1898. $60-75. **Right:** P-134 Wood handle corkscrew with separating tube over shank. Made by Williamson. $50-60.

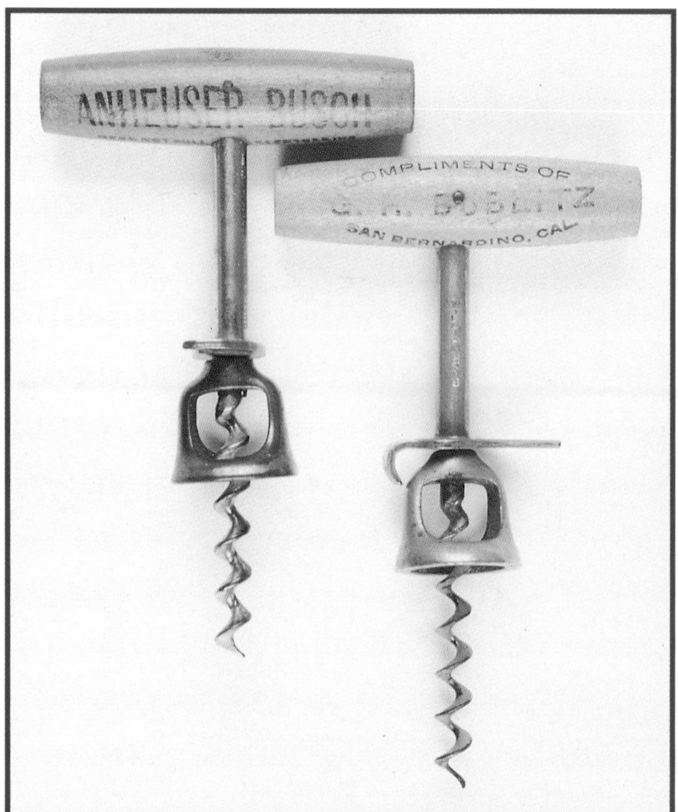

Left: P-82 Wood handle corkscrew with flat wire breaker above cast bell. Marked WILLIAMSON CO., NEWARK, N. J. PATENTED. $15-50. **Right:** P-137 Wood handle corkscrew with wire breaker/cap lifter above cast bell. Marked WILLIAMSON CO., NEWARK, N. J. PATENTED. $25-35.

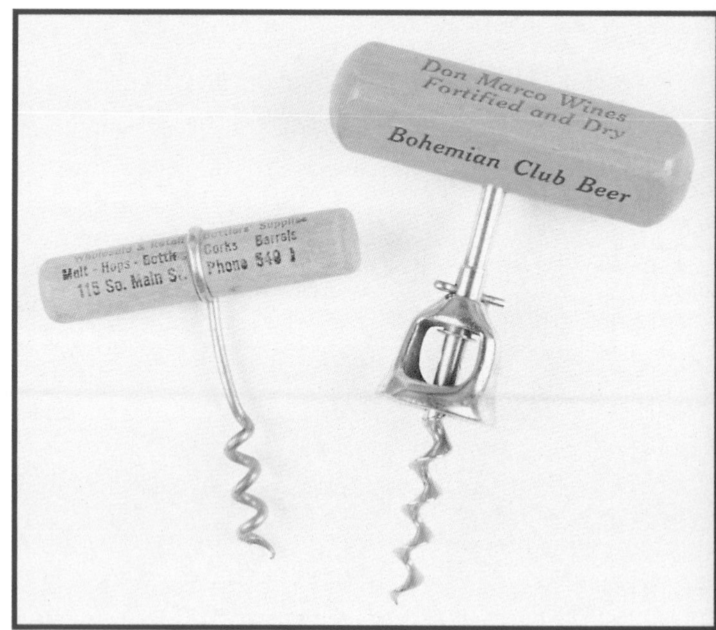

Left: P-95 Wood handle corkscrew with worm wire wrapped around handle. $35-50. **Right:** P-117 Four sided wood handle with cast bell. Made by Williamson Company. $50-60.

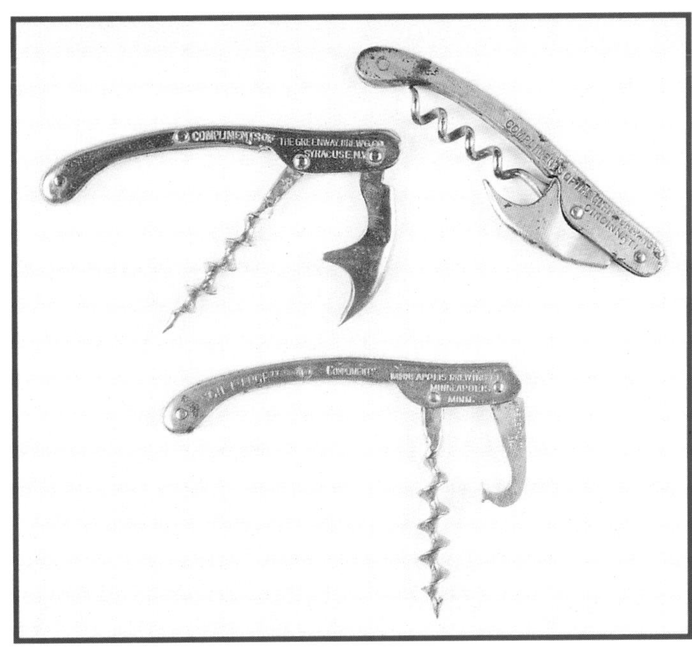

Top pair: P-2 The "Davis" corkscrew. Waiter's Friend type. American patent 455,826 issued to David W. Davis, July 14, 1891. An early Montgomery Ward's catalog says of the Davis: "will fit any bottle and any woman or child can operate it." Web helix and wire helix versions shown. $100-200. **Bottom:** P-70 Waiter's Friend. Corkscrew with cap lifter. Marked THE DETROIT PAT. JULY 10,'94. American patent 522,672 issued to Charles Puddefoot, July 10, 1894. $100-200.

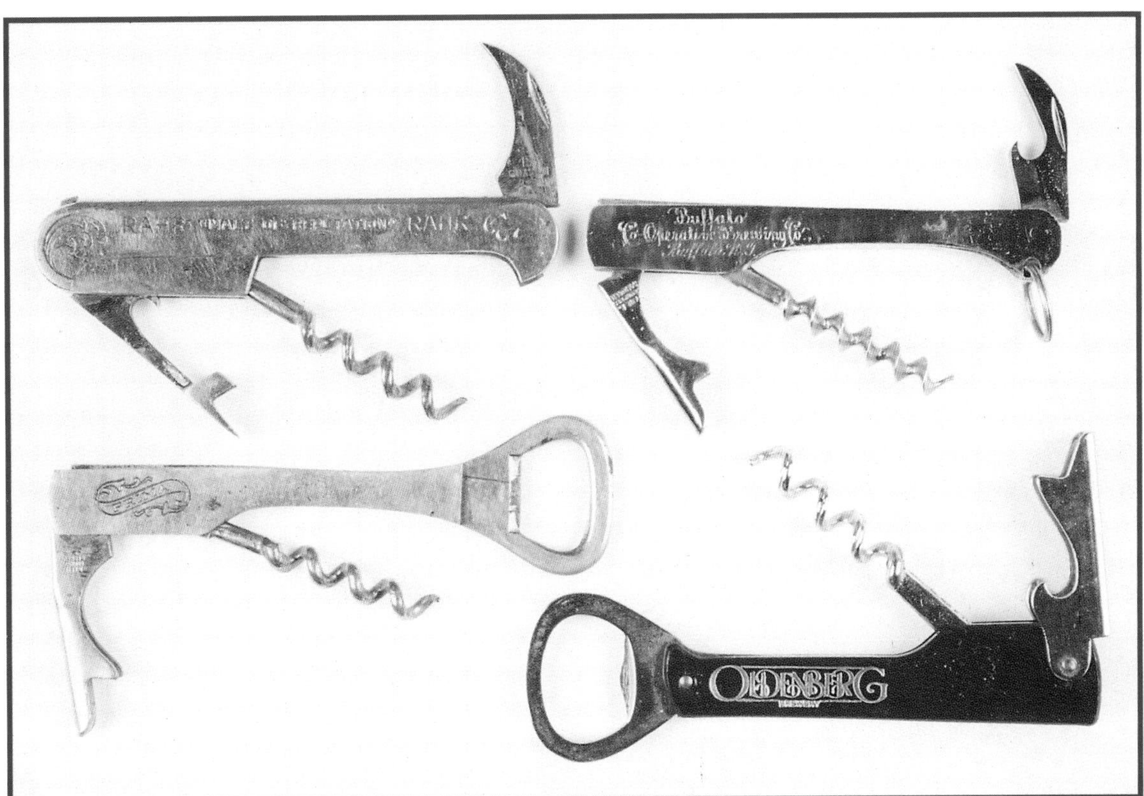

***Above:* Top left:** P-112 Waiter's Friend. Corkscrew with cap lifter on neckstand. Foil cutter. $75-100. **Top right:** P-130 Waiter's Friend. Marked PAUL A. HENCKELS SOLINGEN GERMANY. $125-150. **Bottom left:** P-122 Waiter's Friend. Cap lifter and body one piece. $50-75. **Bottom right:** P-92 Waiter's Friend. Corkscrew with round cap lifter and neckstand with cap lifter. Plastic handle. $5-10.

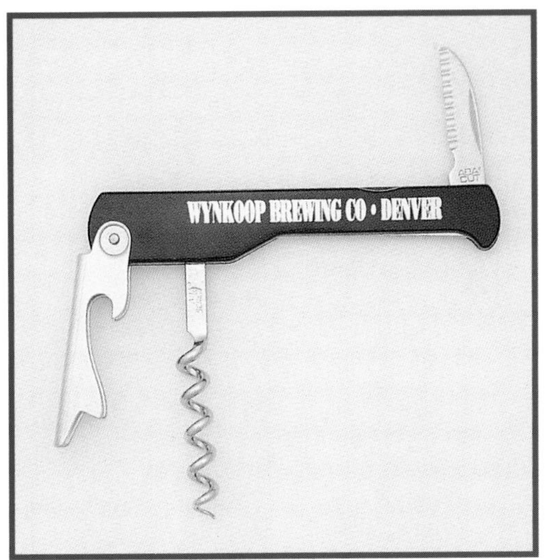

Left: P-147 Modern Waiter's Friend corkscrew, cap lifter, and wire breaker. Marked FRANMARA ITALY. $8-10.

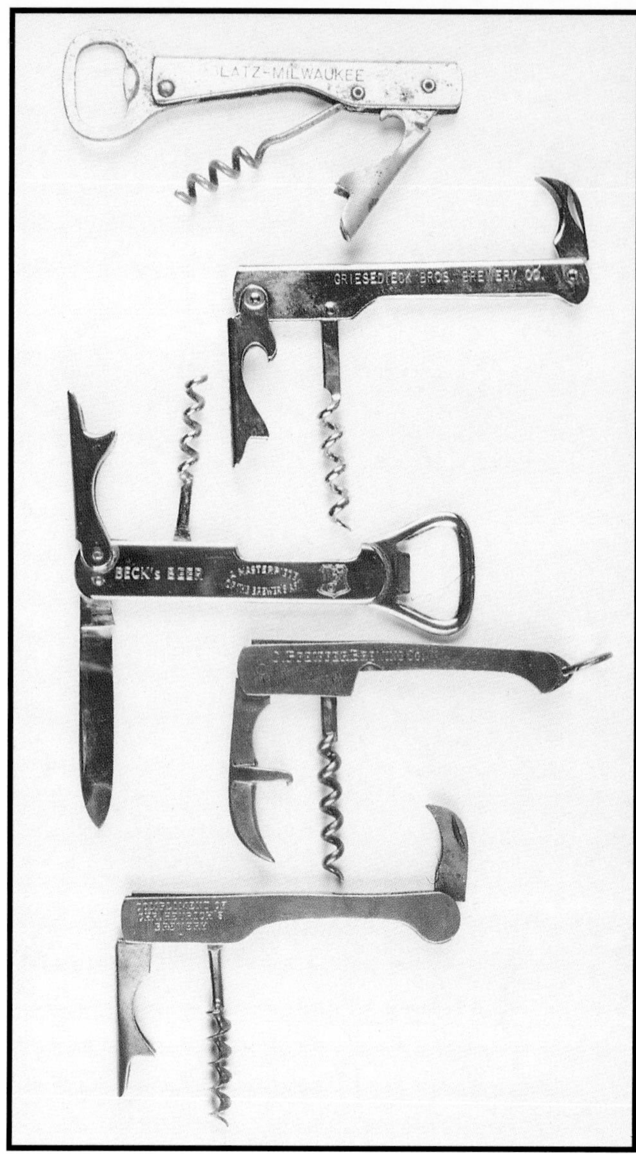

***Right:* Top to bottom:** P-20 The "Bottle-Boy" waiter's friend. Corkscrew and cap lifter. Marked A & J., U. S. A. $20-25. P-21 Waiter's friend with foil cutter opposite neckstand. Made by Universal Cutlery Company, Switzerland. $50-75. P-27 Waiter's Friend. Single knife blade, cap lifter, and corkscrew. $25-35. P-28 Waiter's Friend. Wire breaker/foil cutter and corkscrew. $50-75. P-49 Waiter's Friend. Corkscrew, foil cutter, and neckstand. Marked D. R. PATENT NO. 20815. This is a German patent issued to Karl Wienke in 1882 (An American patent was issued to him in 1883). Wienke is often considered to be the father of the waiter's friend. $75-100.

***Below:* Left side, diagonally:** P-14 Corkscrew knife with master blade having a cap lifter at the base. Marked CAMCO, U. S. A. or KENT, N. Y. C. or COLONIAL, PROV., R. I. $10-30. P-36 Single blade knife with cap lifter at base of blade. Corkscrew on opposite side of blade. $10-30. P-66 Single blade knife with cap lifter at base of blade and corkscrew. Made by Camillus Cutlery Company, Camillus, New York. $10-30. P-15 Single blade corkscrew knife with cap lifter formed in handle. $10-30. **Right side, diagonally:** P-46 Single blade knife with cap lifter formed in handle and corkscrew. Marked THE IDEAL. $10-30. P-129 Knife with master blade, corkscrew, and cap lifter blade with screwdriver tip. Celluloid handles. Marked UTICA CUTLERY CO. UTICA, N.Y. $60-75. P-16 Corkscrew knife with master blade and pen blade. Cap lifter formed in handle. Manufactured by Imperial Knife Company, Providence, Rhode Island. $20-50.

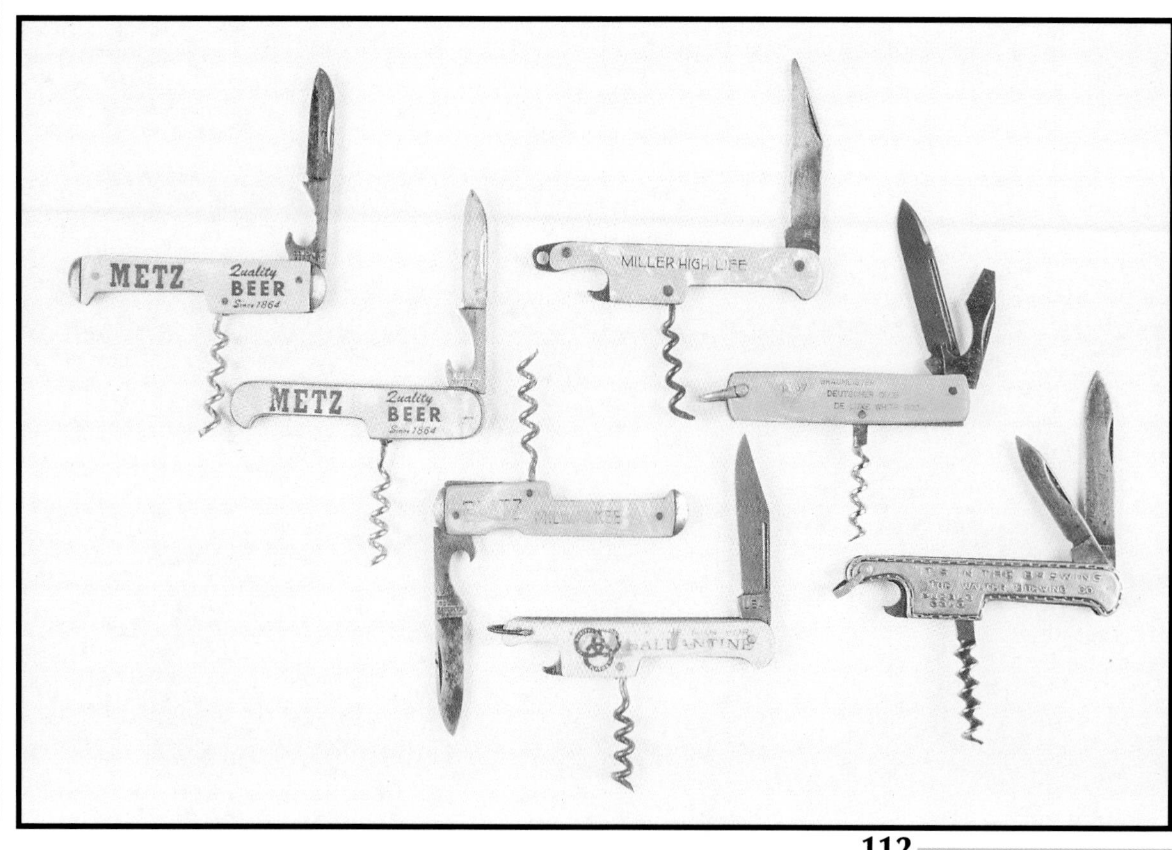

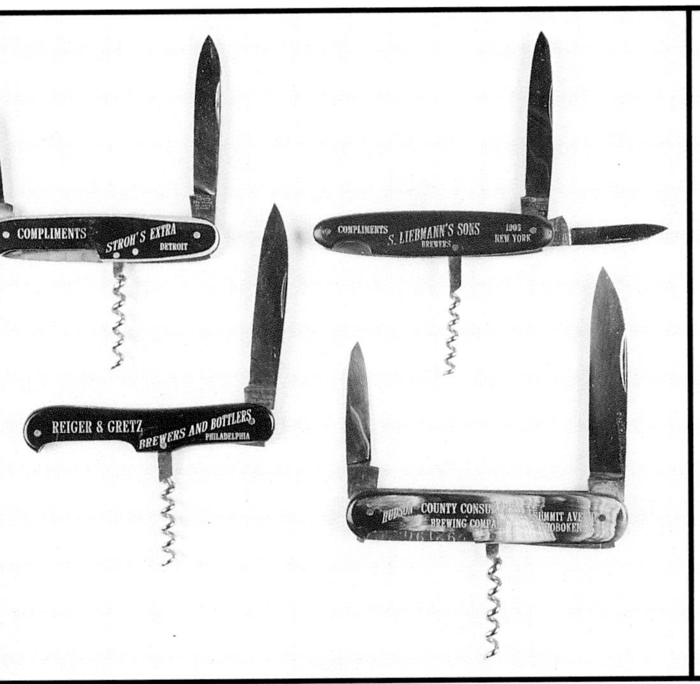

Salesman's sample knives. **Top left:** P-136 Knife with master blade, corkscrew, and pen blade. Celluloid handles. Marked CHRISTIANS SOLINGEN GERMANY/ROSTFREI. $100-150. **Bottom left:** P-142 Knife with master blade and corkscrew marked CHRISTIANS SOLINGEN. $100-150. **Top right:** P-143 Knife with master blade, corkscrew, and pen blade. Marked CHRISTIANS SOLINGEN. $100-150. **Bottom right:** P-144 Knife with master blade, corkscrew, and pen blade. Marked CHRISTIANS SOLINGEN. $100-150.

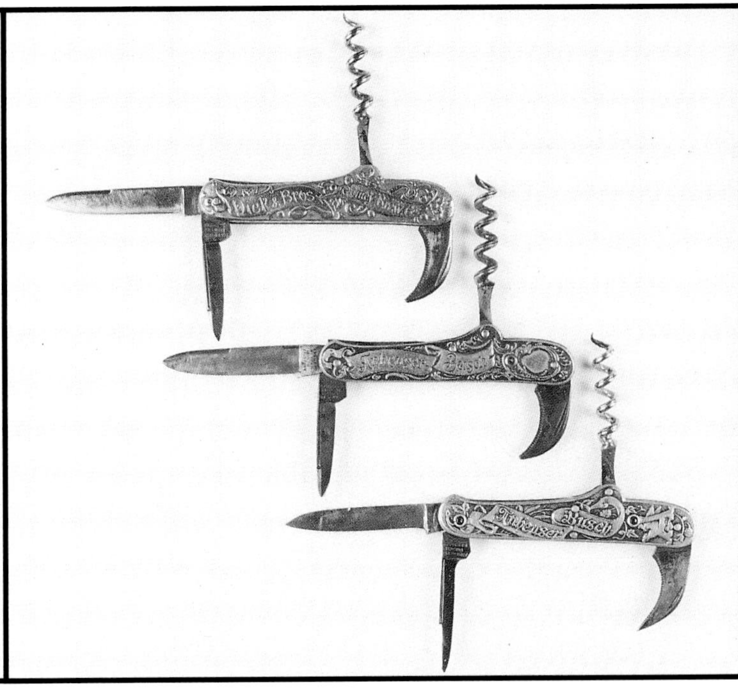

Top: P-26 Two blade knife with wire/foil cutter and corkscrew. $150-250. **Middle:** P-25 Champagne pattern knife with two cutting blades, corkscrew, cap lifter, and wire cutter. Ornate handles. Some have Stanhope (peephole to view photo when held up to light). $150-250. **Bottom:** P-38 Champagne pattern knife with two cutting blades, foil cutter, and corkscrew. Two Stanhopes (peepholes). $250-350.

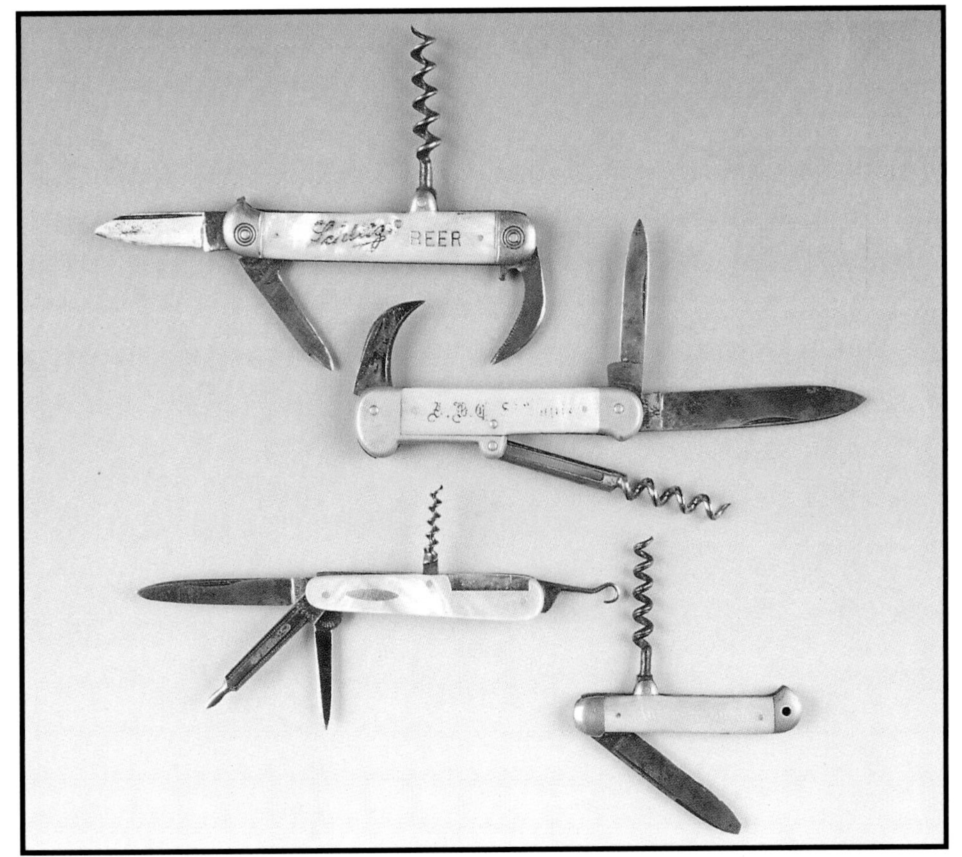

Top to bottom: P-34 Knife with master blade, foil cutter, and corkscrew. Mother-of-pearl handles. Made by Wester Brothers, Solingen, Germany. $125-150. P-40 Knife with sliding corkscrew, wire cutter blade, and two cutting blades. Marked J. A. HENCKELS, SOLINGEN. Mother-of-pearl handles. $300-400. P-110 Knife with two cutting blades, file blade, button hook, and corkscrew. Mother-of-pearl handles. Advertising is on the blade tangs. Made in Germany. $250-300. P-145 Two blade pen knife with corkscrew. Mother-of-pearl handles. Blade marked ANHEUSER-BUSCH, MADE IN GERMANY. $150-250.

Left: P-65 Bottle shape single blade knife. Cap lifter and corkscrew. Steel handles. $150-200. **Right:** P-128 Knife with master blade, corkscrew, and cap lifter formed in handle. $50-60.

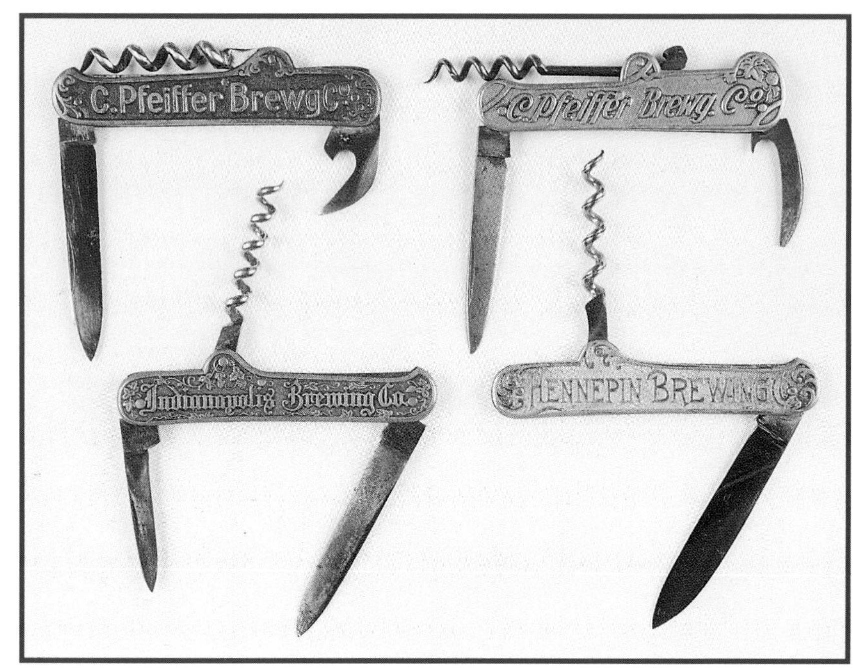

Top left: P-69 Single blade knife with cap lifter and corkscrew. Ornate handles. $125-250. **Top right:** P-108 Knife with master blade, foil cutter/opener blade and sliding/locking corkscrew. Marked M. WORMSER GERMANY. $250-300. **Bottom left:** P-102 Two blade knife with corkscrew. Ornate brass handles. $125-150. **Bottom right:** P-146 Single blade knife with corkscrew. Ornate handles. Marked C. F. KAYSER, SOLINGEN. $150-175.

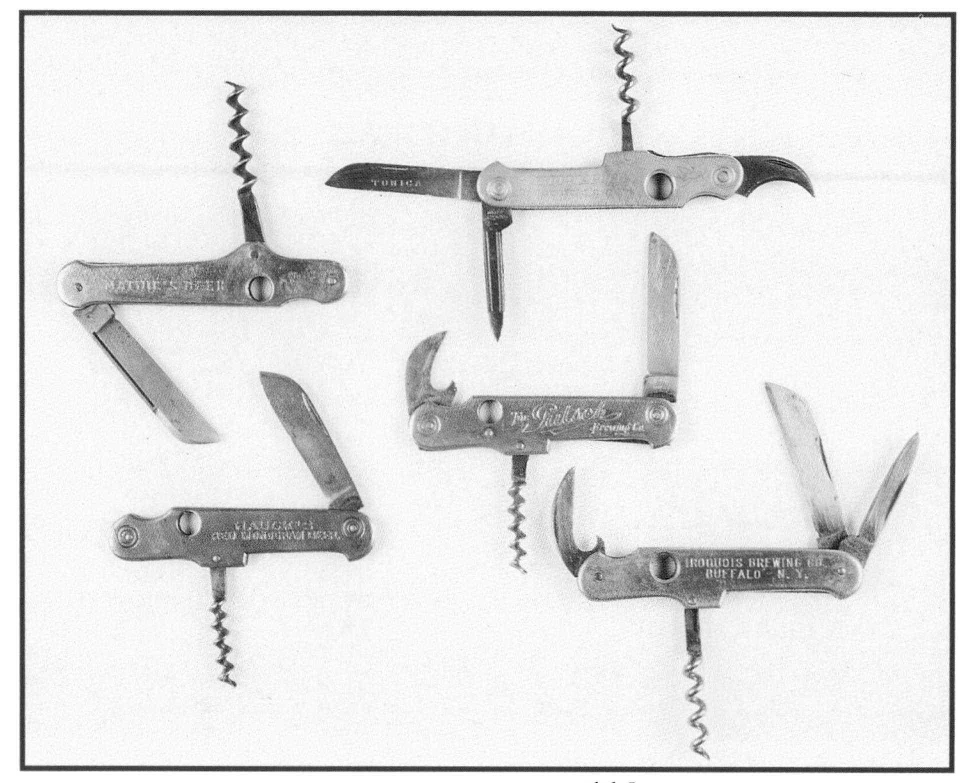

Left side, top to bottom: P-48 Master blade cuts cigar tips. Web helix corkscrew. Plain metal handles. $125-150. P-101 Knife with cigar cutter blade and corkscrew. $125-150. **Right side, top to bottom:** P-29 Knife with master blade used to cut cigars, file blade, wire/foil cutter, and corkscrew. Plain metal handles. $125-150. P-96 Knife with cigar cutter blade, cap lifter, and corkscrew. Marked H. KESCHNER SOLINGEN GERMANY. $150-200. P-125 Knife with two cutter blades, cap lifter blade, cigar cutter, and corkscrew. Marked A. W. WADSWORTH & SONS N. Y. $125-150.

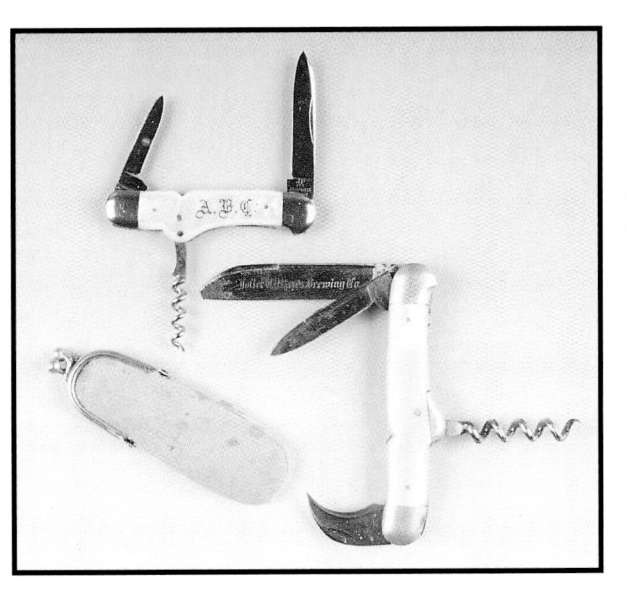

Left: P-107 Two blade knife with corkscrew with leather case shown below. Mother-of-pearl handles. Marked J. A. HENCKELS, SOLINGEN. $250-300. **Right:** P-121 Knife with two cutting blades, wire breaker. Mother-of-pearl handles. Marked D. PERES SOLINGEN GERMANY MAGNETIC CUTLERY CO. $150-200.

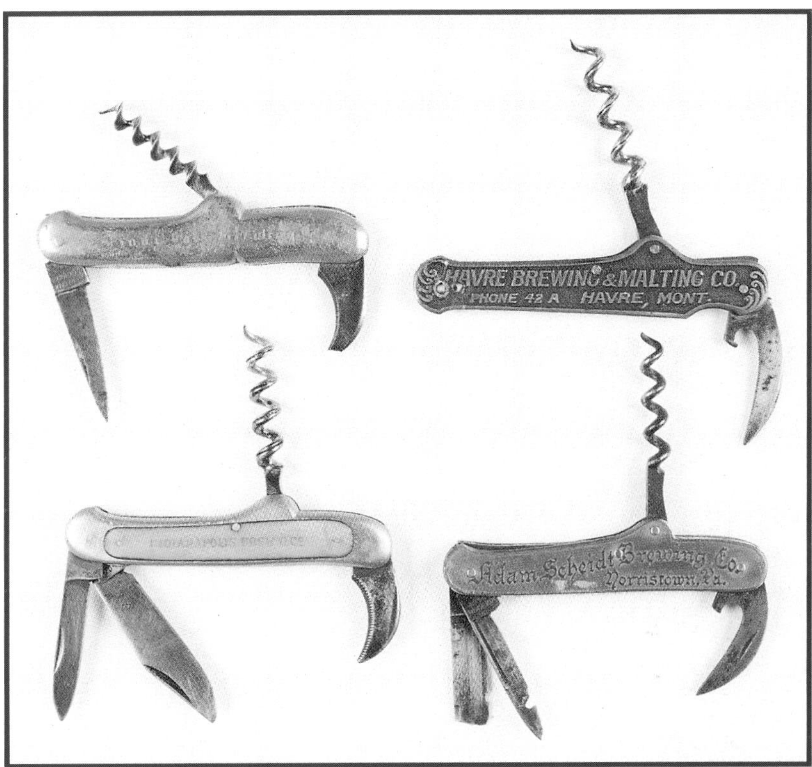

Top left: P-39 Knife with master blade, foil cutter, and corkscrew. Plain metal handles. $150-200. **Top right:** P-123 Corkscrew with cap lifter/foil cutter blade. $150-200. **Bottom left:** P-105 Knife with master blade, foil cutter, and corkscrew. Celluloid insert handles. Made in Germany. $125-150. **Bottom right:** P-120 Knife with two cutter blades, cap lifter blade, and corkscrew. Ornate handles. Marked D. HERDER OHLIGS GERMANY or J. A. HENCKELS GERMANY. $100-150.

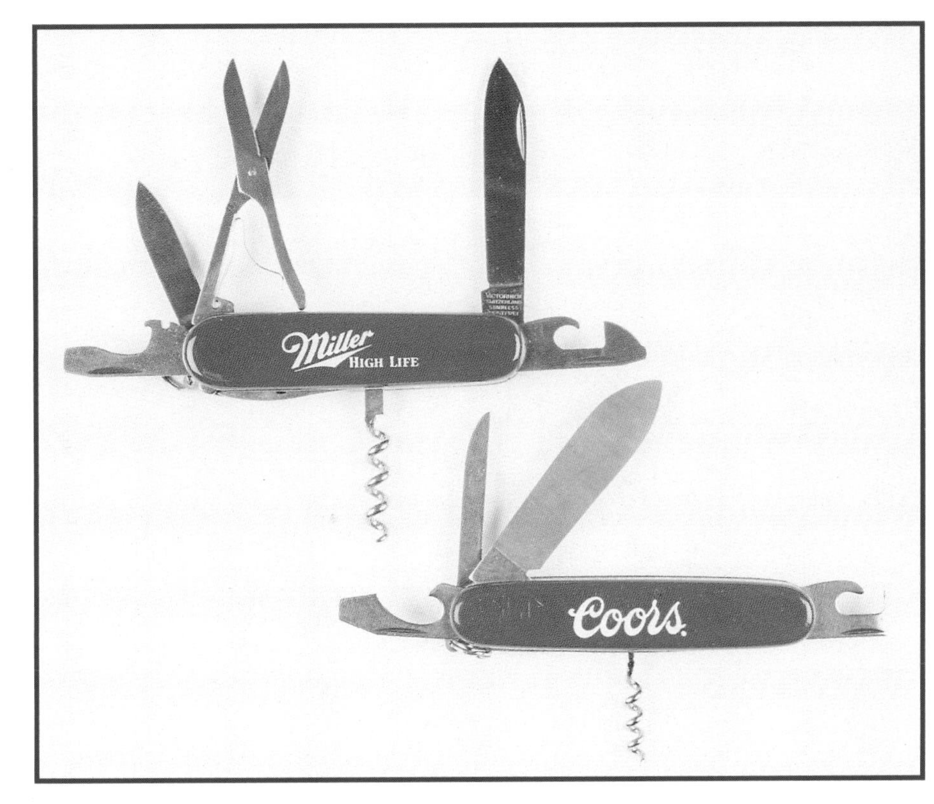

Top: P-104 Swiss Army knife with corkscrew. Marked VICTORINOX SWITZERLAND STAINLESS ROSTFREI, OFFICIER SUISSE. $20-25. **Bottom:** P-113 Sportsman's knife with corkscrew. Manufactured by Colonial Knife Company, Providence, Rhode Island. $20-30.

P-30 Bar mounted corkscrew. Crank handle to insert worm into cork; lift lever to extract. American patent 452,625 issued to Edwin Walker, May 19, 1891. $500-700.

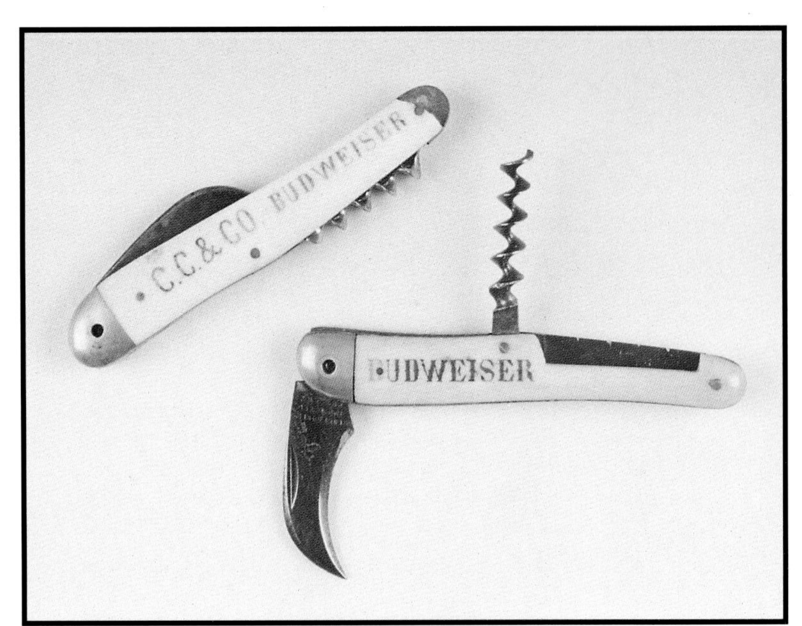

P-71 Corkscrew with foil cutter blade. Mother-of-pearl or bone handles. With and without Stanhope. Made by Henry Boker's Improved Cutlery Co. $250-300.

P-35 Bar mount corkscrew. The "Quick & Easy" by Walker. Screws to bar top. There are four different versions with four different design marks inside the handle: 1893 DESIGN, 1895 DESIGN, 1896 DESIGN, and 1897 DESIGN. Crank handle to insert screw into cork and to extract. $500-650.

Left: P-64 "Champion" bar mounted corkscrew. Manufactured by the Arcade Manufacturing Company of Freeport, Illinois. American patent 589,574 and design patent 25,607 issued to Michael Redlinger, September 7, 1897 and June 9, 1896. Hold bottle in clamp and move handle from back to front to insert worm and extract cork. $500-700.

Below left: P-83 "Yankee No. 7" bar mounted corkscrew. Secure bottle in clamps. Then pull handle forward to enter cork and to extract in one motion. American patent 1,058,361 issued to Raymond B. Gilchrist, April 8, 1913. $500-700.

Below: P-84 Bar mounted corkscrew. Rotate top handle to enter cork. Pull lever to extract cork. American patent 377,790 issued to Edwin Walker, February 14, 1888. $600-800.

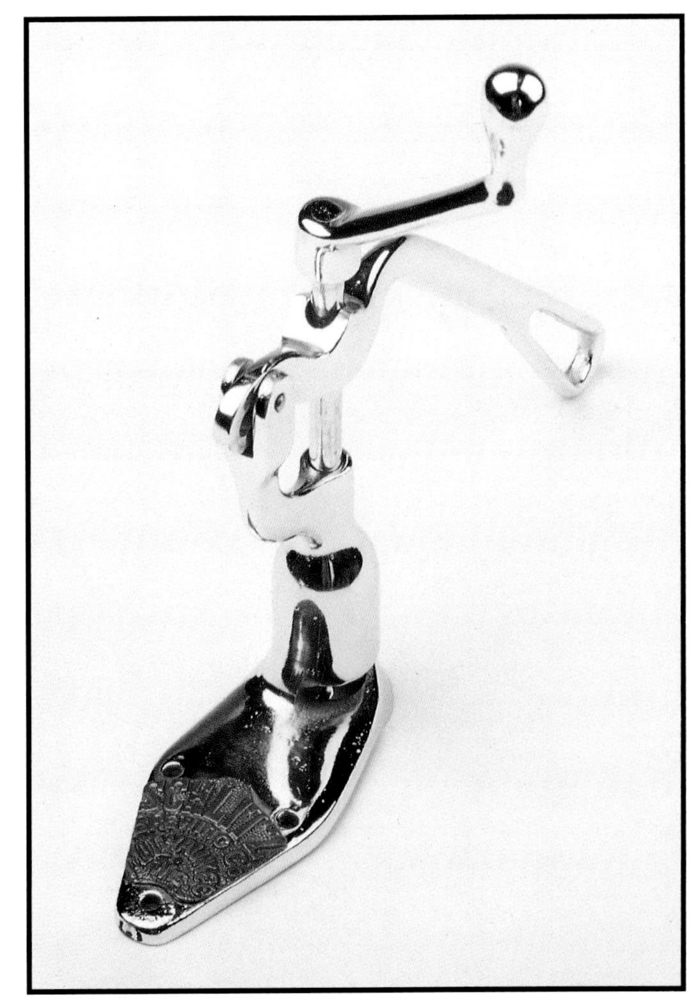

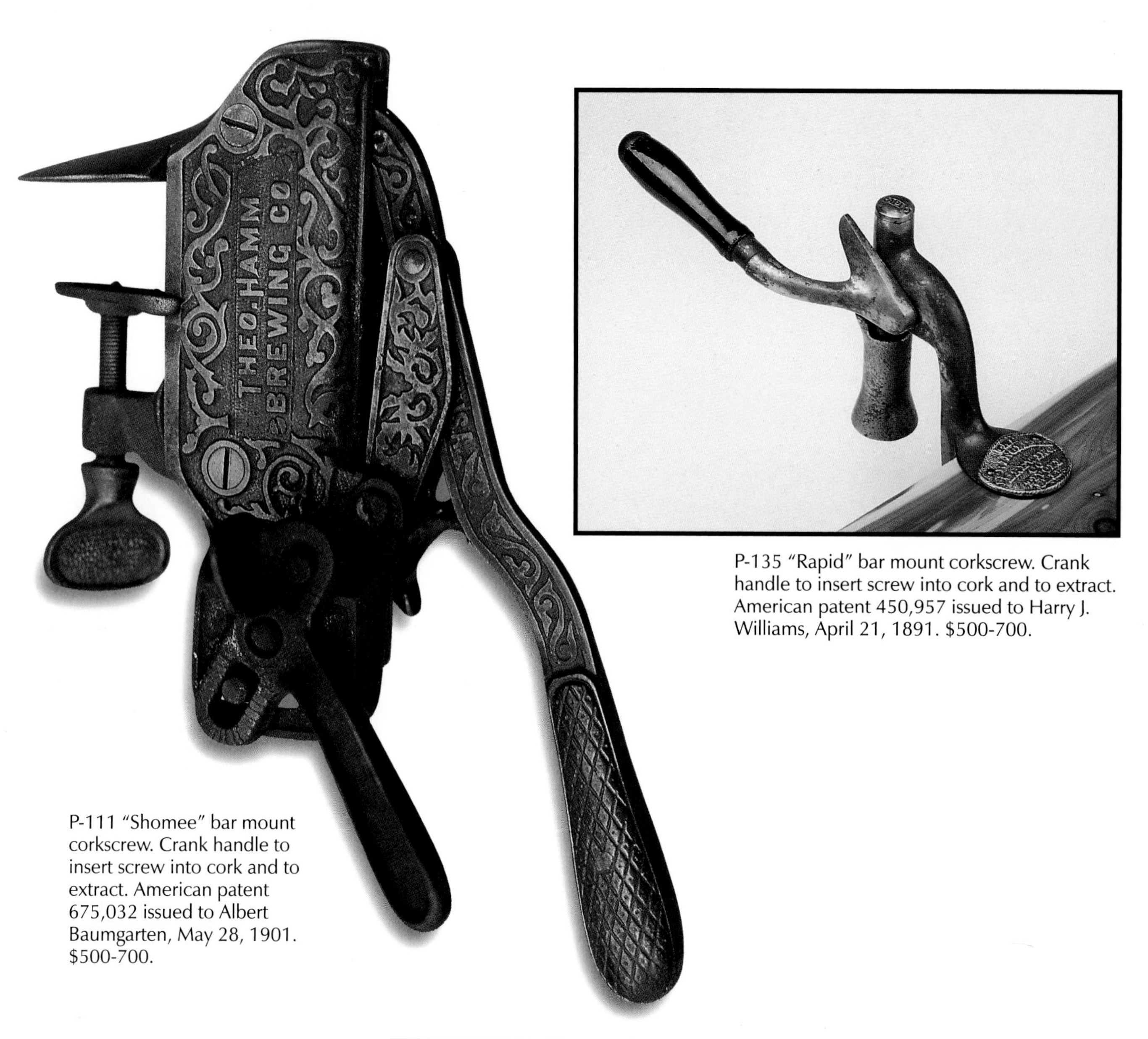

P-135 "Rapid" bar mount corkscrew. Crank handle to insert screw into cork and to extract. American patent 450,957 issued to Harry J. Williams, April 21, 1891. $500-700.

P-111 "Shomee" bar mount corkscrew. Crank handle to insert screw into cork and to extract. American patent 675,032 issued to Albert Baumgarten, May 28, 1901. $500-700.

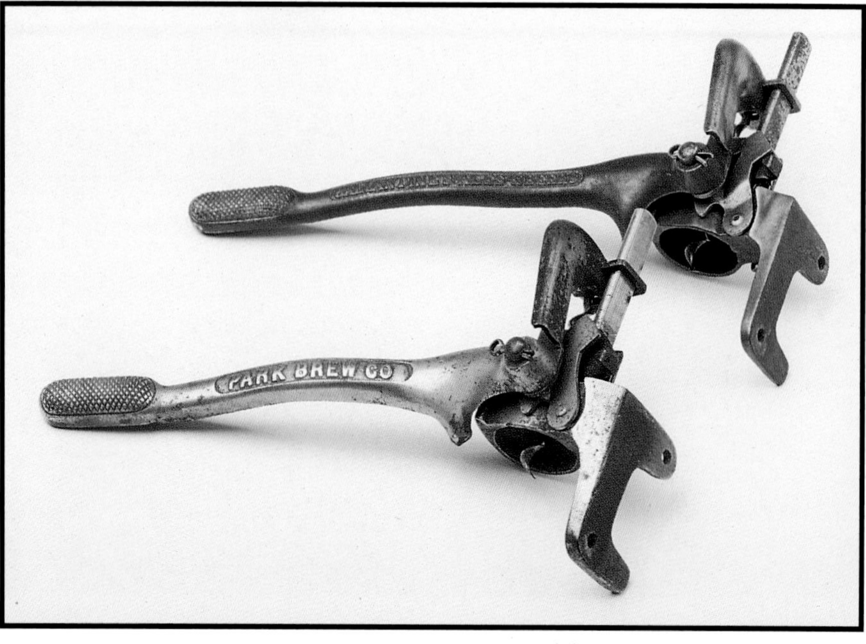

P-45 "Yankee No. 1" wall mount corkscrew. American patent 857,992 issued to Raymond B. Gilchrist, June 25, 1907. The Gilchrist Co. of Newark, New Jersey, marketed the Yankee wall mounted cork puller as "a household necessity." Advertising copy says: "Should be in every home. Don't let any woman struggle with a corkscrew to open tightly corked catsup, olive, pickle, medicine on any other bottle. The Yankee is screwed against any upright surface: Icebox, Sideboard, Door Frame or Wall. It's always there. No hunting for a corkscrew, always ready to draw the tightest cork from any bottle." $500-700.

Modern Openers

In 1962, the Pittsburgh Brewing Company introduced a can with a pull tab that required no "opener." That and the introduction of the twist off bottle cap brought on the age of the Modern Openers. Some still have the conventional cap lifter, but most are for use with push tabs, flip tabs, pull tabs, pop tops, twist off caps, and combinations of any or all of these. This colorful lot of plastic and metal openers has added a whole new arena for the opener collectors. Advertising specialty houses are having a field day soliciting the large national brewers and the explosive microbrewery market. As a result, a stop by any brewery gift shop will yield an opener addition to a collection. They are still cheap, but they will be tomorrow's antiques.

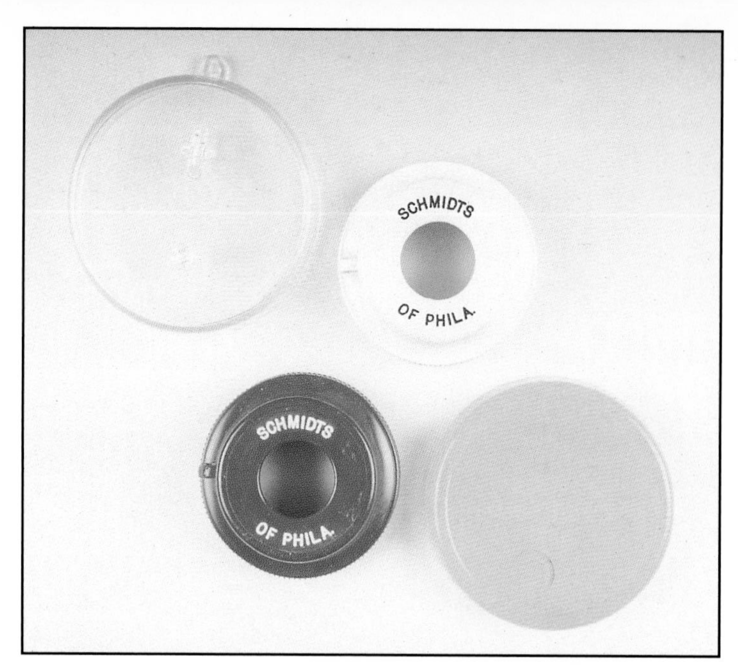

Top left: Q-2 Opener for pop top cans with key chain tab. 2 7/8" wide and 5/8" high. Two examples shown. $3-5. **Middle pair:** Q-1 Push tab can opener. Place on top of can, press down, and tabs are depressed. 3 3/4". $3-5. **Bottom right:** Q-4 Opener for pop top cans with no key chain tab. 2 3/4" wide and 1/2" high. $8-10.

Top left: Q-3 Plastic grip with metal cap lifter insert with magnet. Marked PAT.PEND LIFT HERE. American design patent 231,313 issued to G. John Heelan, April 16, 1974. 2 1/2". $2-4. **Top right:** Q-21 Twist off cap remover. Marked PAT PEND MADE IN CHINA. 3 1/4". $2-4. **Bottom left:** Q-11 Plastic grip with metal cap lifter/twist off cap remover insert. Magnet holds cap. Marked INVENTEX RO 1983, MADE IN CANADA. 3". $2-4. **Bottom right:** Q-53 Twist off cap remover with refrigerator magnet. Marked CAP 2 CAP WORLDWIDE PATENTED L. A. CA U. S. A. SLIDE BOTTLE CAP IN EASY OPENER HERE, PRESS DOWN TO OPEN (with picture of hand and bottle) SLIDE BOTTLE CAP IN EASY OPENER HERE, THIS SAMPLE IS FOR SHOW PURPOSE ONLY LOGO USED IS OWNED SOLELY BY THE COMPANY SHOWN AND IS A REGISTERED TRADE MARK. SAMPLE NOT FOR SALE. 2 3/4". $2-4.

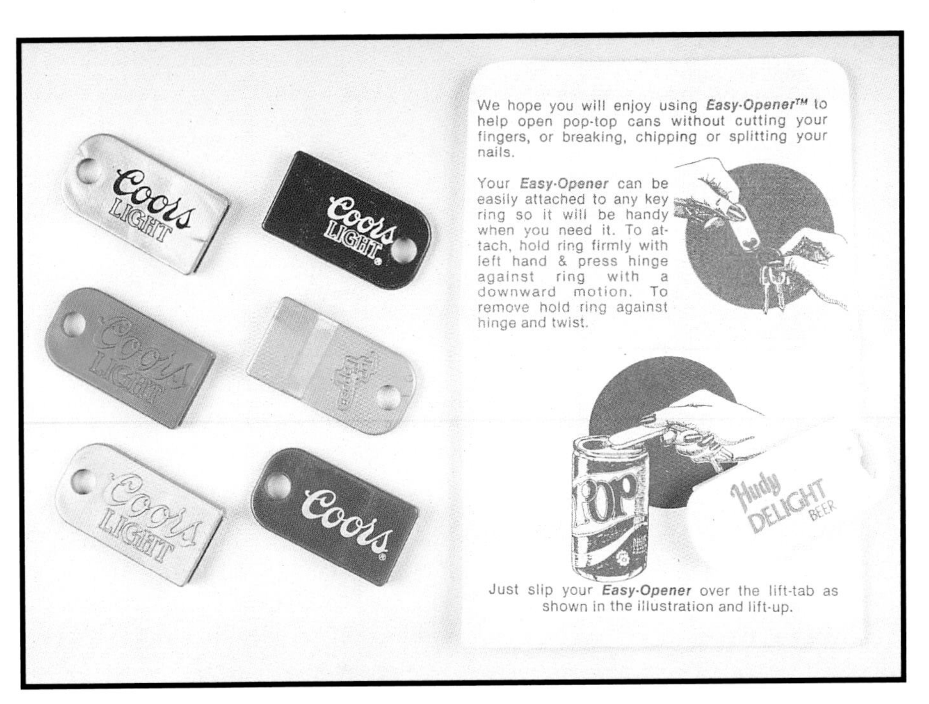

Left group of six: Q-5 Top popper for can tabs. Plastic. 1 5/8". $1-2. **Right:** Q-6 Top popper for can tabs. Plastic. 2 1/4". $1-2.

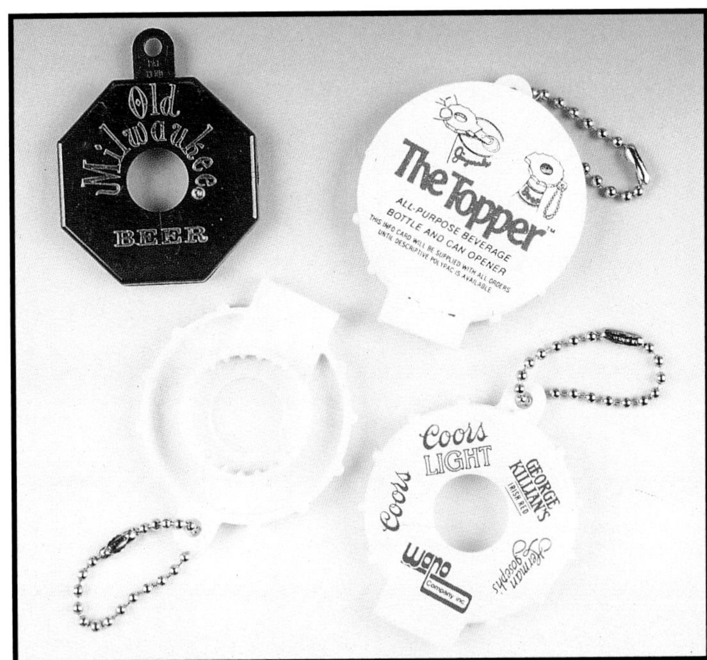

Top left: Q-7 "Beer Twist" twist off bottle cap opener. Octagon grip. Marked MEGA MANUFACTURING PO BOX 9051 SAN DIEGO, CA. 92109 BEER TWIST USA. 2". $2-4. **Other three:** Q-10 "The Topper." Twist off cap remover and top popper for can tabs. 2 1/8". $2-4. *A commemorative example celebrates "Pearl Brewing Co., 1886 - 1986."*

Left pair: Q-8 Latch hook key ring with cap lifter. May be marked MADE IN ITALY ACTION LINE U.S.A. PAT. NO. 4,414,865 or ACTION LINE U.S.A. PAT.PEND. 3 1/8". $3-5. **Middle:** Q-19 Latch hook key ring with cap lifter. Marked TAIWAN. 3 5/8". $3-5. **Right:** Q-22 Cap lifter with tab lifter. Round shield on handle. Marked 1985 L B LUCAS PAT PENDING. 2 5/8". $3-5.

Q-9 "Magigrip" opener. Grips and twists off caps. 3 1/8" to 5" diameter. $2-4. **Bottom right:** Q-51 Twist off cap remover. 2 5/8". $3-5. *Produced in a series with bowl number and score result for each Super Bowl on football handle. Advertisement for Lite Beer from Miller Brewing Company.*

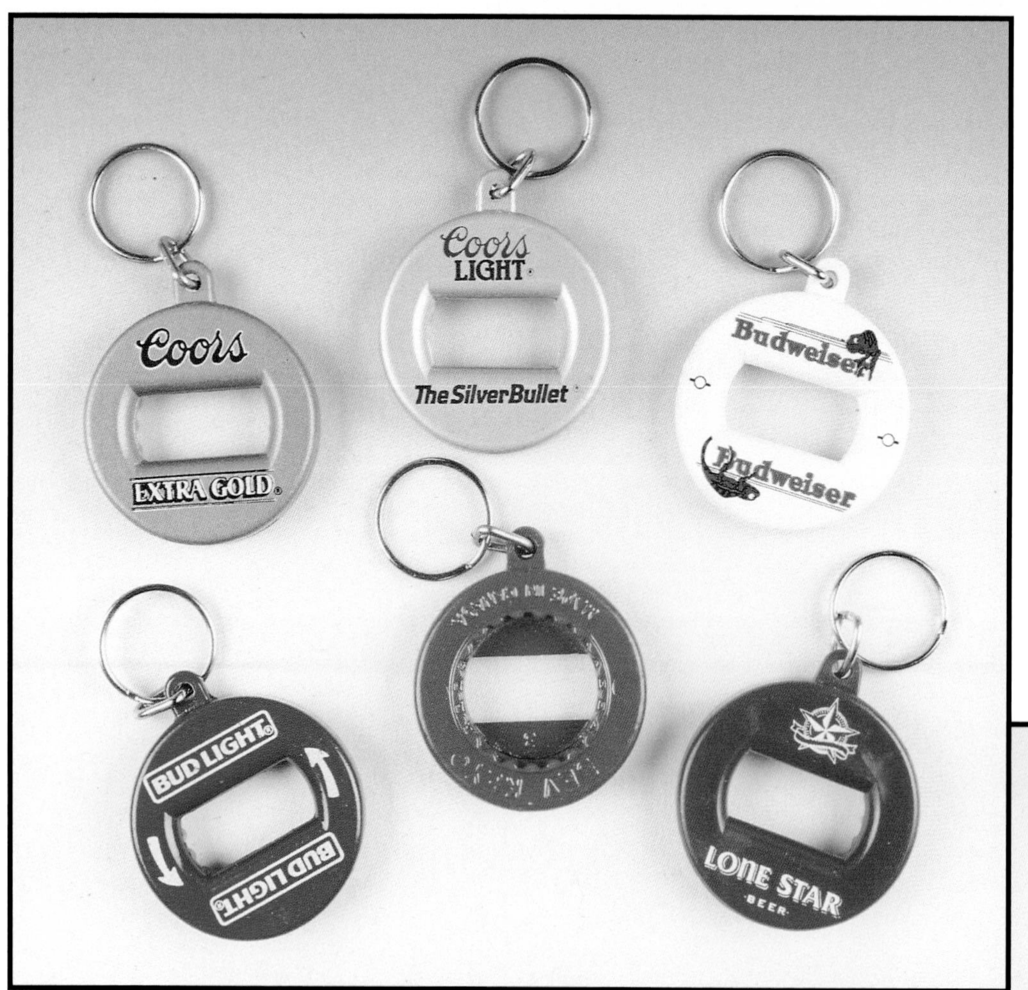

Five similar: Q-12 Cap lifter with twist off cap remover in back and tab lifter slot in the side. All metal. Marked MADE IN CANADA BEV KEY. 1 7/8". $3-5. *This type was used as a souvenir of the 1993 New Orleans "canvention" of the Beer Can Collectors of America.* **Bottom right:** Q-44 Plastic with twist off cap remover in back and no tab lifter slot. Marked MADE IN CHINA or ENDURO USA. 2". $3-5.

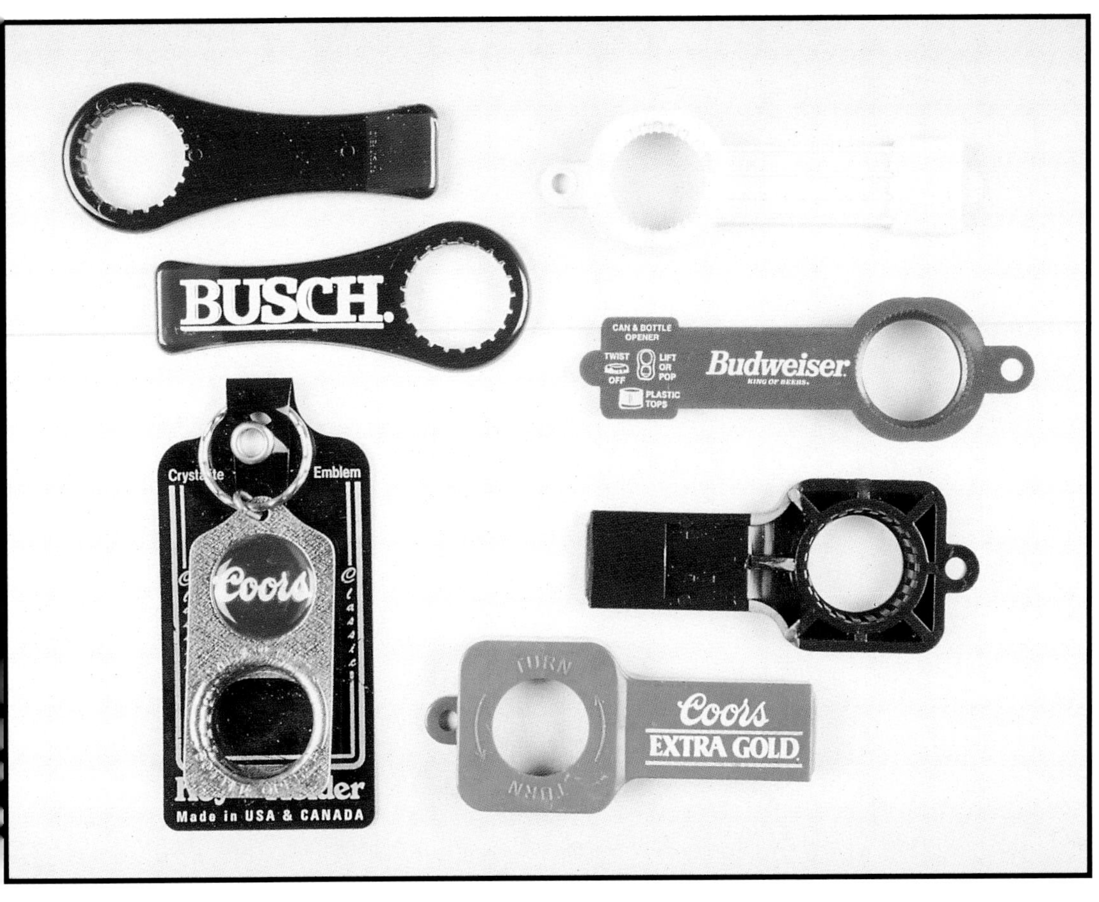

Top left pair: Q-15 Twist off cap remover and tab lifter. Marked PATENT PEND. 4″. $2-4. **Bottom left:** Q-28 Twist off cap remover. Marked TWIST TO LIFT BOTTLE OPENER. 3″. $3-5. **Top right pair:** Q-35 "E-Z 3-Way Opener" twist off cap remover and tab lifter. Marked E-Z 3-WAY OPENER PAT.NO. D 316,362 U.S.A. PAT.NO. 4,723,465 CANADIAN PATENT NO. 1,260,207. 4 3/4″. $3-5. **Bottom right pair:** Q-39 Twist off cap remover. Marked PAT.NO. 321 TURN. 4 1/4″. $3-5.

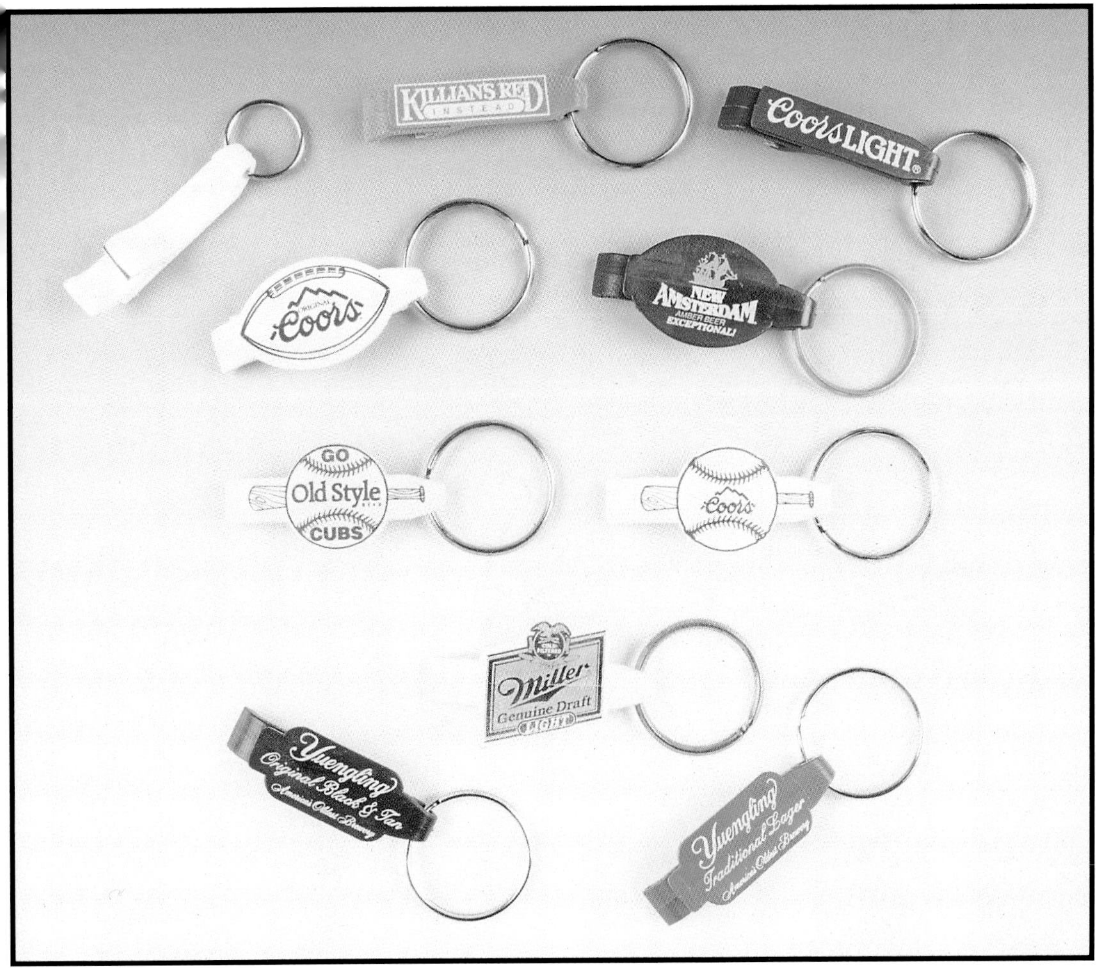

Top three: Q-18 Cap lifter with tab lifter. Marked EVANS MFG SA CA 92704 PAT 4,864,898. 2 5/8″. $2-4. **Next two:** Q-31 Cap and tab lifter. Football theme. 2 5/8″. $2-4. **Middle pair:** Q-32 Cap and tab lifter. Used with various sports themes including bowling, basketball, pool, volleyball, baseball, and golf. 2 5/8″. $2-4. **Next one:** Q-33 Cap and tab lifter. Miller Beer label on handle. 2 5/8″. $2-4. **Bottom pair:** Q-36 Cap lifter. 2 1/2″. $2-4.

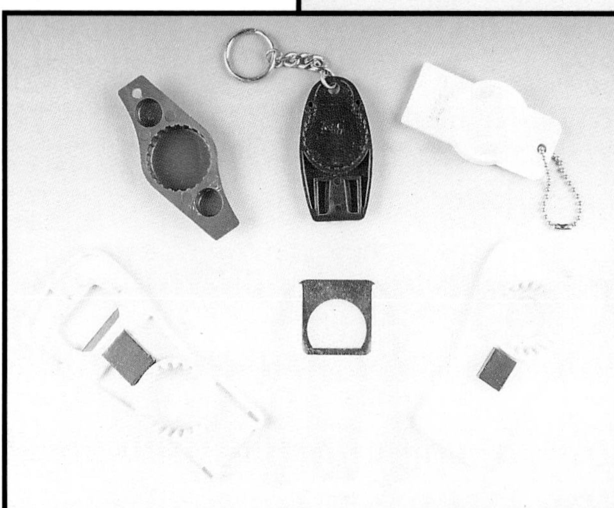

Top, left to right: Q-16 Twist off cap remover. 3 1/8". $2-4. Q-17 "Lifter-Twister" twist off cap remover, tab lifter, and key ring. Marked LOTTERY LIFTER-TWISTER PAT. PEND. LIFT TAB. 2 1/2". $3-5. Q-24 Twist off cap remover. Marked TAB OPENER TOGMASTER. 2 3/4". $2-4.
Bottom, left to right: Q-23 Cap lifter and twist off cap remover with magnet. Marked PATENT PEND. USA. 4 3/4". $3-5. Q-30 Cap lifter with twist off cap remover slot. Marked MADE IN W. GERMANY DBP PATENTED PLASTOLAN. 3 3/4". $2-4. Q-20 "Tab Grabber." Twist off cap remover and tab lifter with magnet. Marked COPYRIGHT THE RON HOFFMAN GROUP, INC. 1983 PATENTED. 3 7/8". $3-5.

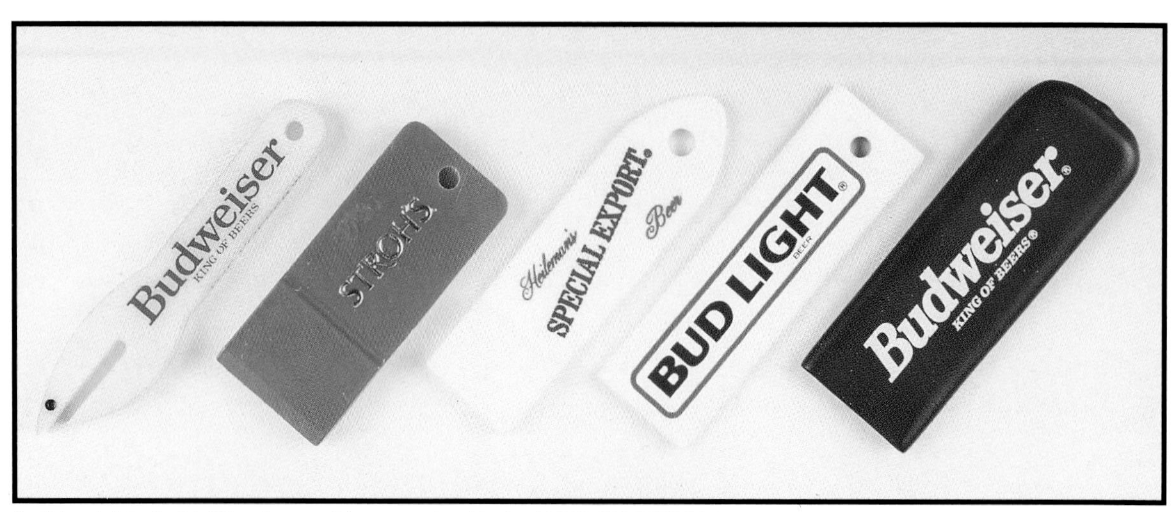

Left to right: Q-14 "Top Popper" for can tabs. Marked TOP POPPER TM COPYRIGHT ANDERSON-KROEGER 1980 PAT PEND. ANDERSON-KROEGER 1981. 3 3/8". $8-10. Q-26 Tab lifter with magnet. Marked CHINA. 2 1/2". $2-4. Q-43 "Save A Nail" tab lifter. Marked SAVE A NAIL. 3 1/8". $2-4. Q-50 Tab lifter. Metal. 3". $2-4. Q-52 Tab lifter. 3". $2-4.

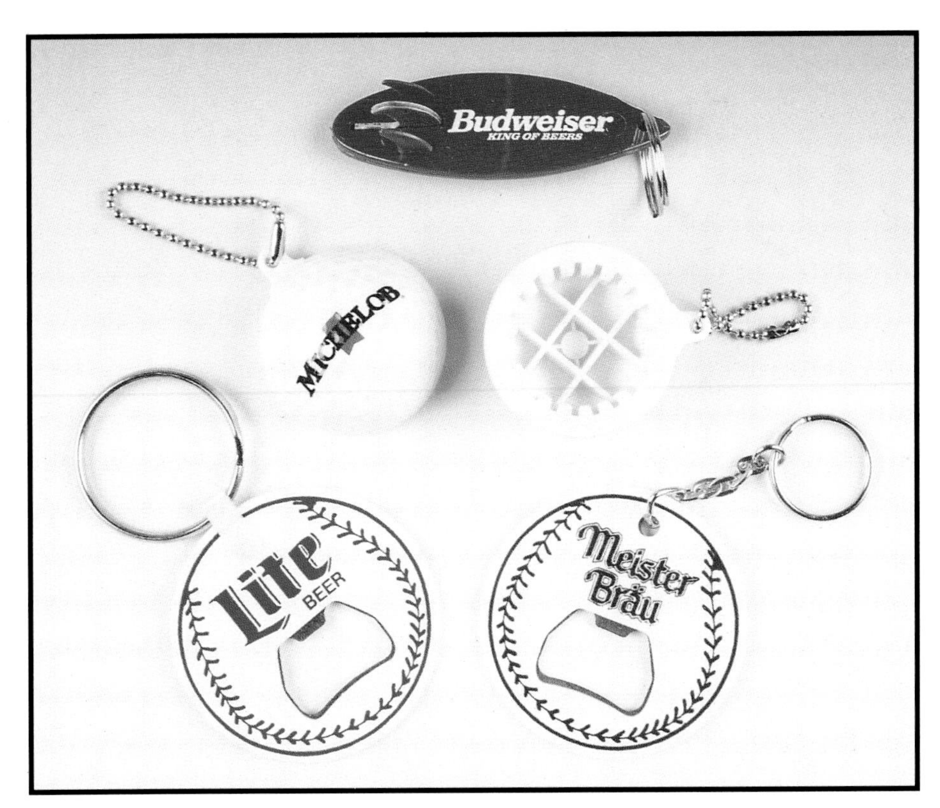

Top: Q-27 Cap lifter. Surfboard shape. Marked MADE IN TAIWAN. 3". $2-4. **Middle pair:** Q-45 Half golf ball twist off cap remover. Marked JSW ENT. 1 5/8" $2-4. **Bottom pair:** Q-46 Cap lifter with baseball design. 2 3/8". $2-4.

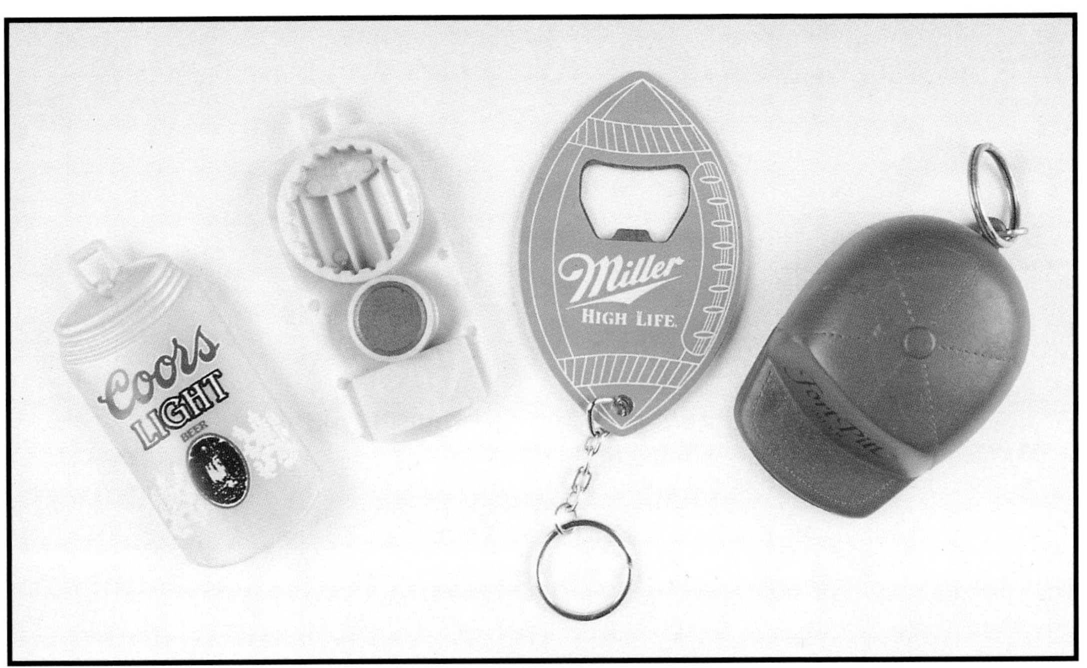

Left pair: Q-29 Twist off cap remover with magnet. Beer can shape. Marked MADE IN CHINA. 3 1/4". Top and bottom shown. $3-5. **Middle:** Q-48 Cap lifter. Football shape. 3 1/2". $2-4. **Right:** Q-49 Twist off cap remover and tab lifter. Baseball cap shape. Marked U.S.A. PAT.PEND. 12. 3 1/8". $5-8.

Left: Q-25 "Pop-A-Tab" finger lift cap remover. Marked POP-A-TAB TM PAT PEND. 1 1/8". $5-8. **Middle:** Q-34 "Li'l Buddy" twist off cap remover. Plastic. Marked LI'L BUDDY PAT. PEND. TAB OPENER. 1 5/8". $5-8. **Right:** Q-42 Can pry opener. Marked OPENER TO OPEN CAN, PRY WITH EDGE PAT NOS. 5,110,002 5,125,525 MADE IN U.S.A. PAT NOS. D. 312,785 5,054,640. 2 3/8". $5-8.

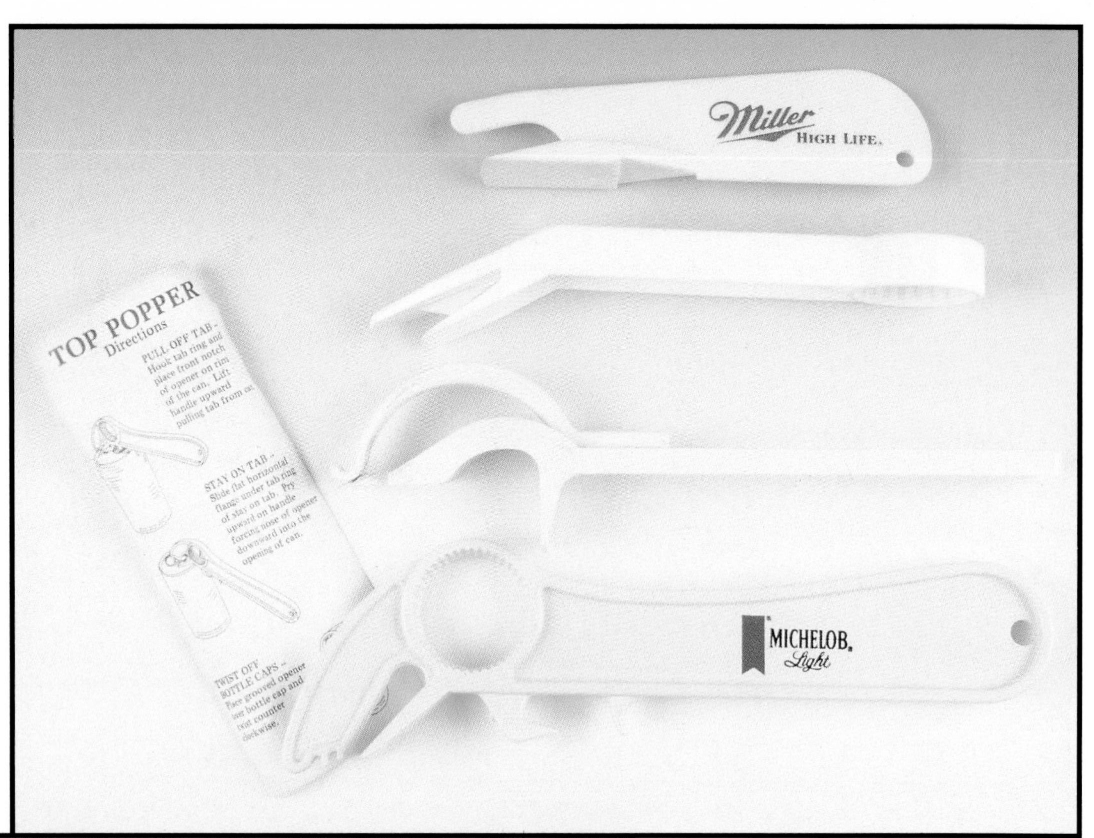

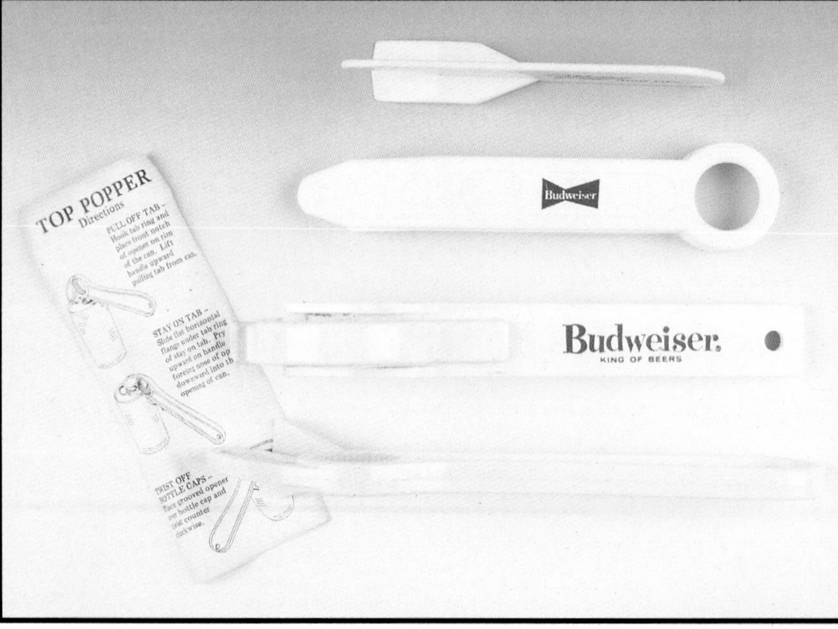

Top to bottom: Q-41 "Pop-A-Top" tab lifter. Marked POP-A-TOP (picture of how to use) ADTREND U.S.A. 4 3/4". $5-8. Q-59 Twist off cap remover and tab lifter. Marked TABMASTER INC. 92041 NO. T-100 PAT. PENDING. 6". $8-10. Q-47 Top popper for can tabs. Marked PATENT PENDING. 7". $5-8. Q-37 "Top Popper" twist off cap remover and cap/tab lifter. 7 1/2". $10-12.

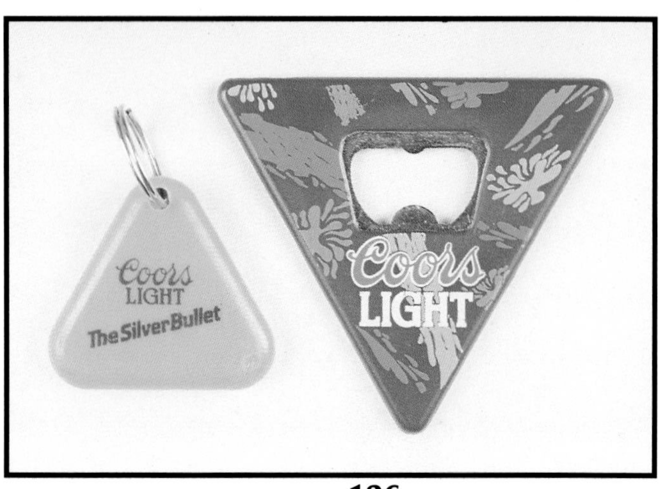

Left: Q-38 Twist off cap remover. Triangular. Marked A GOOD SPORT NO. PAT. PENDING U.S.A. 1 7/8". $3-5. **Right:** Q-54 Triangular shape cap lifter. Made in China. 3 1/2". $3-5.

Left: Q-40 Twist off cap remover, cap lifter, prier, magnet, and ball-point pen. Marked KOREA PAT. 36977 TWIST PRY LIFT. 6". $5-8. **Middle:** Q-60 Cap lifter, prier, magnet, and ball-point pen. Marked MADE IN KOREA. 5 5/8". $8-10. **Right:** Q-61 Cap lifter. Plastic over metal handle. Marked SCARONI ITALY. 4 1/2". $3-5.

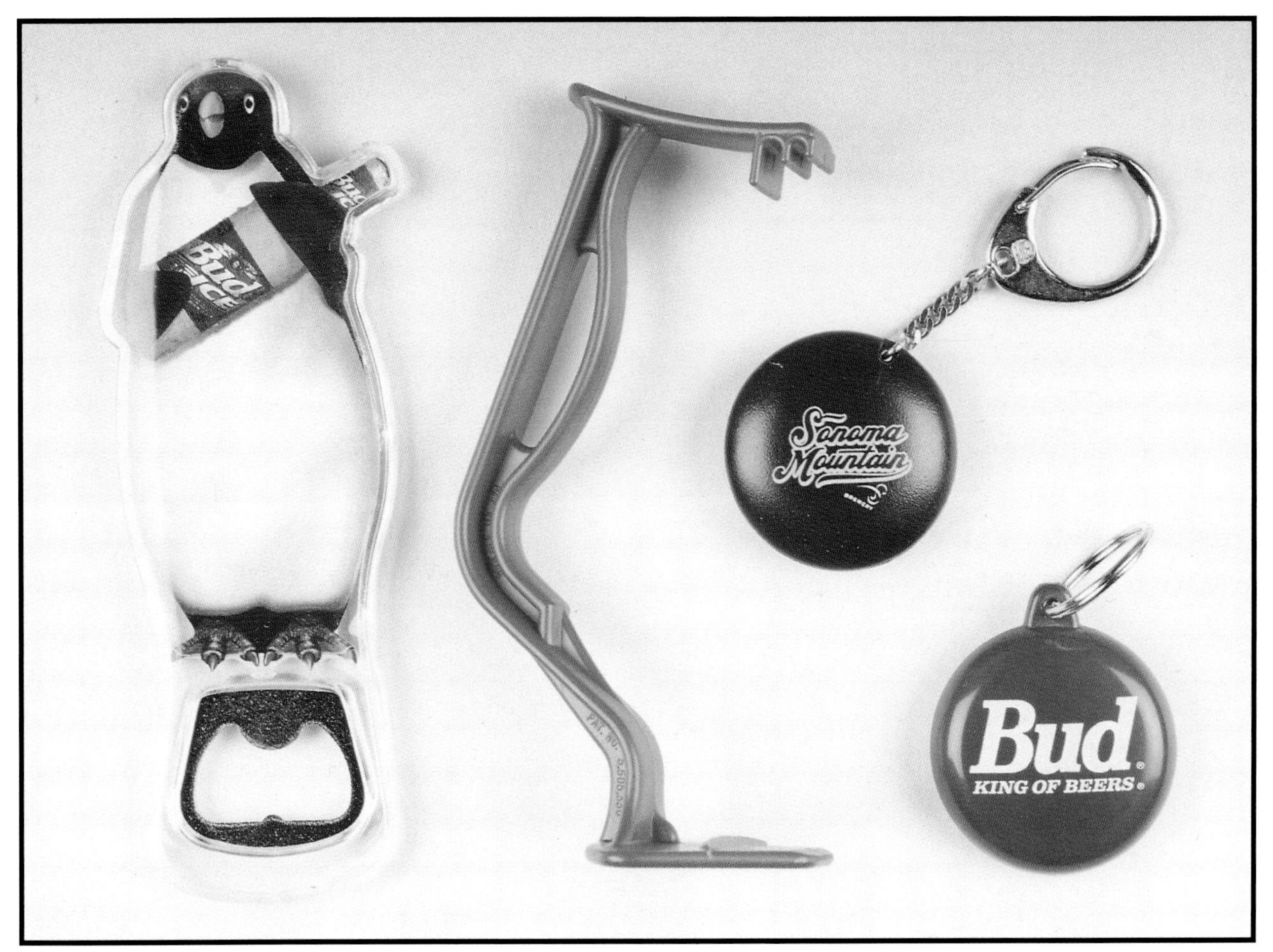

Left to right: Q-55 Plastic encased penguin cap lifter. Made in Taiwan. 5 7/8". $5-8. Q-56 Tab lifter beer can handle. 5". $3-5. Q-57 Round plastic disk cap lifter with picture of "how to use opener." Marked HAACHEN GERMANY DESIGN INT. PAT. 1 3/4". $2-4. Q-58 Round plastic disk twist off cap lifter. 1 3/4". $2-4.

Figurals (3-D)

The original intention of this category was to include openers depicting a person or other three-dimensional figures. By stretching your imagination, you might be able to see a bird, or perhaps a fish in type R-14. The trivet and auto-whisk are probably misplaced, but we'll keep them here lest we disrupt our alphanumeric system!

Note: Type R-1 through R-4 date from 1954 to 1955. They advertise either Iroquois Beverage Corp. or International Breweries, Inc. The name change took place during this period.

Left: R-1 Figural Indian. Red plastic with metal cap lifter on back. 4 1/2". $10-15. **Right four:** R-2 Tin and steel with flat back. Marked ROTARY BUFFALO. In bronze, chrome, and chocolate tones. 3 3/8". $20-30.

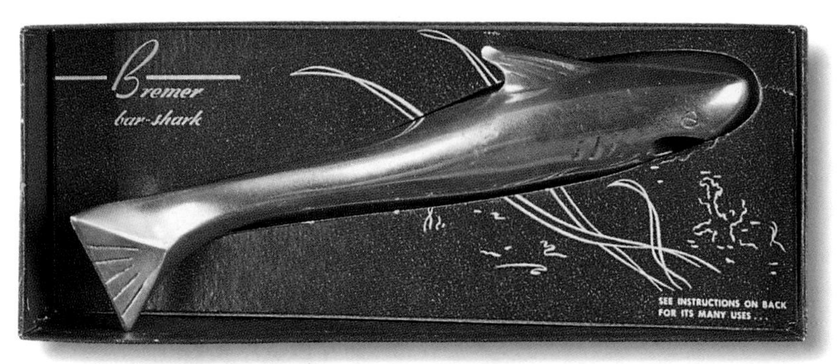

Above: **Left:** R-3 Free standing three-dimensional Indian. Feathers form cap lifter. Produced in Aluminum, brass, and magnesium. 4 3/4". $20-50. **Right:** R-4 Free standing three-dimensional Indian. Feathers form cap lifter. Top hinged to expose lighter. 5". $250-300.

Left: R-5 "Bar Shark" opener. Aluminum. Made by Bremmer Manufacturing Company, Milwaukee, Wisconsin. Packaging says "Body - used as an ice cracker; Nose - used as a muddler; Mouth - used for opening bottles; Tail - for removing push collar caps; Fin - for breaking cellophane wrap on neck of bottle." 7". $100-125.

R-6 Wall mounted cap lifter. Painted cast iron or brass. Depicts brewmaster with mug of beer. Cast Iron opener marked COPYRIGHT MR LANC, PA. 6 1/4". $300-400 (brass); $400-600 (painted).

Above: R-7 Cast iron pretzel cap lifter. 2 7/8". $60-75.

Left: R-8 Wall mounted cap lifter. Painted cast iron. Depicts coal miner with mug of beer. Marked COPYRIGHT MR 1933. 6 3/4". $800-1000.

R-9 Baseball cap with cap lifter insert in base. Marked LOYAL PROD NYC BOTTLE OPENER. 3 3/4". $100-150. *Found painted "Rheingold 1963 Award."* R-13 Bronze hawk cap lifter. Beak is the cap lifter. A limited edition (50) by the Mendocino Brewing Company of Hopland, California, was called "Spirit of the Hawk." 2 1/2". $20-25. R-16 Football helmet with cap lifter insert in bottom. Marked TAIWAN. 2 3/4". $5-8.

R-10 Figural dog cap lifter. Brass. American design patent 79,877 issued to M. D. Avillar, November 12, 1929. 5". $100-150. R-11 Figural parrot cap lifter and corkscrew. Chrome plated. American design patent 78,544 issued to M. D. Avillar, May 21, 1929. 5 1/2". $75-100.

Above: R-12 "Bar Trivet" cast iron opener. Made in Japan, Pat. No. 134873 & 499791. 4 3/4". $20-25. R-15 "Auto-Whis-Kit." Cap lifter with brush and coin-key holder. Marked AUTO-WHIS-KIT PAT 218387 WHIT RAMSEUR, N.C. 2 3/4". $10-15.

Left: R-14 Cap lifter and jar lid remover. 8". $75-100.

Top, left to right: R-17 Shark cap lifter. Plastic. 3 1/4". $5-8. R-18 Copper kettle with cap lifter insert in bottom. Made in Germany. 1 7/8". *Currently available from Franconia Brewing Company, Pennsylvania for $10.* R-22 Frog twist off cap lifter. Rubber. 3". $3-5. **Bottom, left to right:** R-23 Lizard cap lifter. Plastic. 3 3/8". $3-5. R-25 Bat cap lifter. Plastic. Marked CHINA. 3" $3-5. **(pair)** R-26 Football cap lifter. Plastic. Marked CHINA. 2 1/2". $3-5.

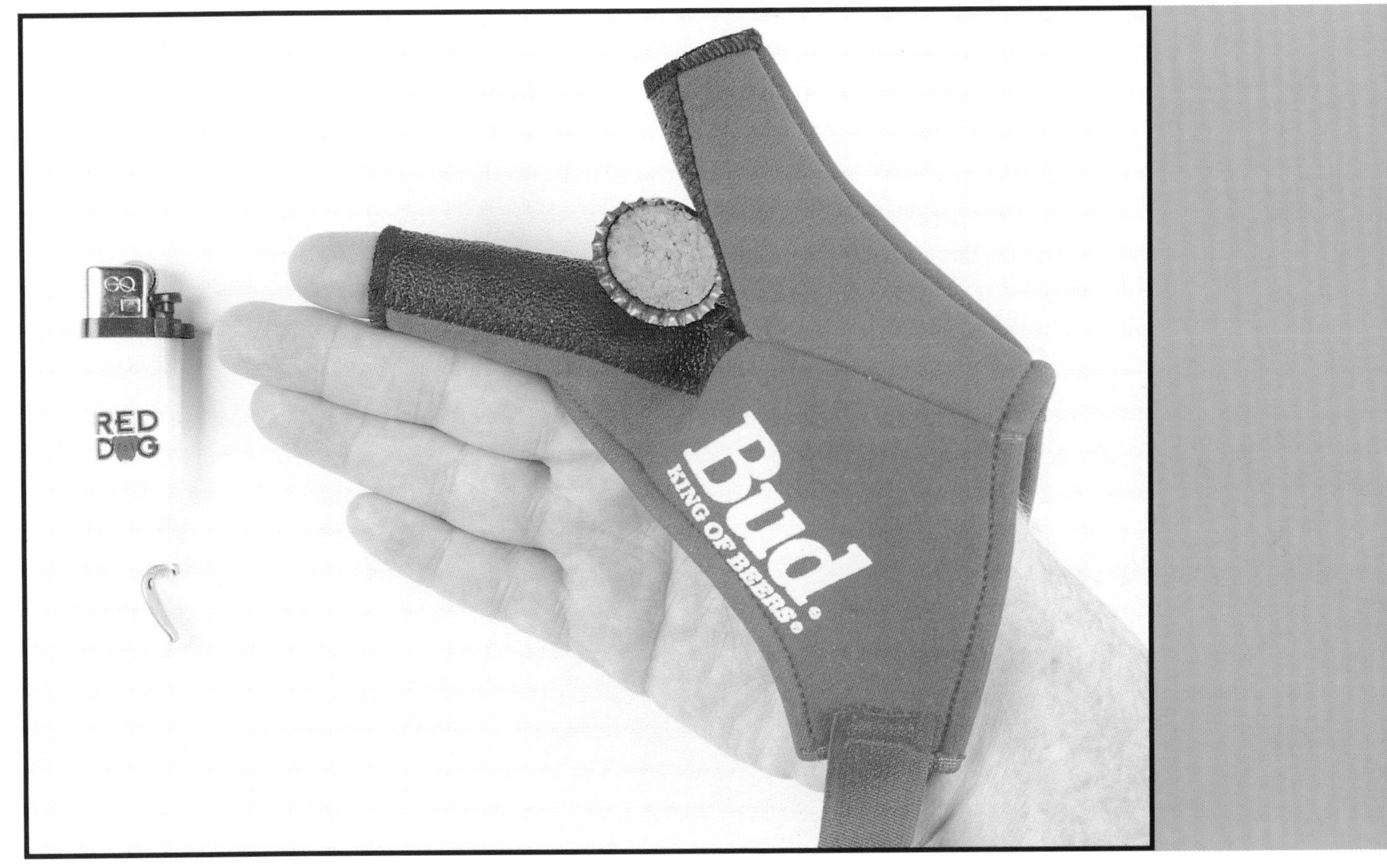

R-19 Cap lifter in shape of lighter. Plastic. 3 5/8". $2-3. R-21 Twist off cap lifter in shape of glove. Called BUD BARTENDER GLOVE. Latex. 7". $5-8.

R-20 Cap lifter in shape of guitar. Pewter. Made for the 1998 debut of Hard Rock Beer. 4". $25-30. R-24 Cap lifter in shape of snowboard. Made in China. Marked W B FIVE ENTERPRISES PAT DES 3851. Steel. 3 1/2". $3-5.

Werner Martinmaas' 1959 patent best describes this category with this amusing excerpt: "Containers such as bottles and cans which are provided with a crown cap are a great source of annoyance on accounts of the frequency with which a person finds himself in possession of several containers of a beverage, but no opener for removing their crown caps. A person is particularly likely to find himself in this predicament while fishing, hunting, or picnicking. The resulting efforts to remove the crown caps from the containers are likely to cause damage to other articles with which removal of the caps is attempted, or serious injury to mouth or teeth if one is foolish enough to attempt to remove caps in that manner."

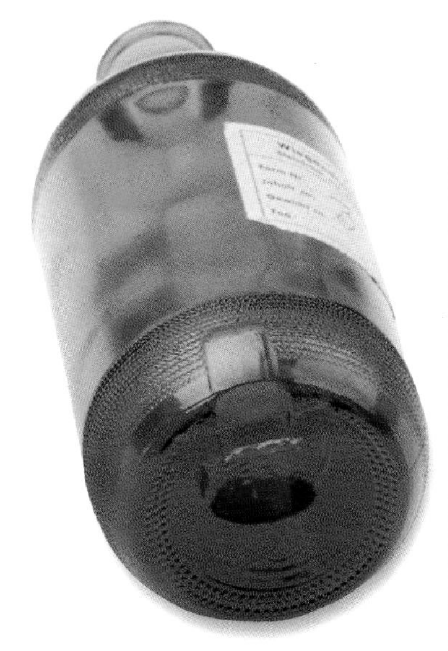

S-1 Bottle with opener formed in glass in the bottom (one would need two bottles to open one). Produced by Fairmount Glass Company, Indianapolis, Indiana. American patent 2,992,574 issued to Werner Martinmaas, March 18, 1959. 5 3/4". $25-35.

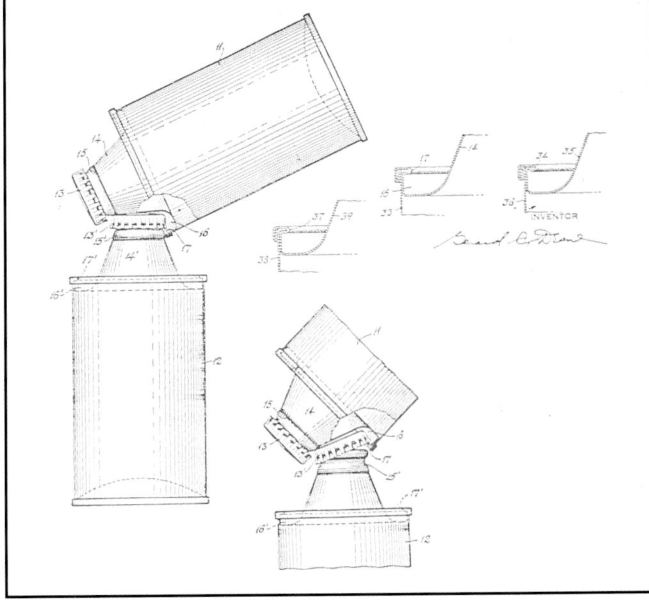

S-2 Cone top can with shallow top to allow another can to be opened. American patent 2,322,843 issued to Gerald Deane, New York, June 29, 1943. $50-100.

S-3 Glass mug with twist off cap remover formed in bottom. 5". $10-20.

Tap Handle Openers

Tap handle openers are simply a regular beer tap handle with an opener attached. In the early 80s, Anheuser-Busch labeled four different tap handle openers as part of their "Clydesdale Collection."

T-2 Tap handle with cap lifter. Various shapes and sizes depending on tap handle attached. $15-25.

T-1 Tap handle with cap lifter and can piercer. Various shapes and sizes depending on tap handle attached. $20-25.

Convention Openers

The organization "Just for Openers" was founded in 1979. Since then, meetings have been held annually at locations from coast to coast. In 1984, convention hosts Gary Deachman and Bill McKienzie introduced a limited edition of openers commemorating the sixth annual meeting. Several other members followed suit resulting in a mini opener collecting category. Convention openers in other categories include 1986 (A-32), 1988 (P-21), 1993 (P-98), and 1995 (M-54).

Left: X-1 1984 "Just for Openers" convention souvenir cap lifter. 5 1/2". $50-60. **Right:** X-2 1985 "Just for Openers" convention souvenir corkscrew. 5". $30-40.

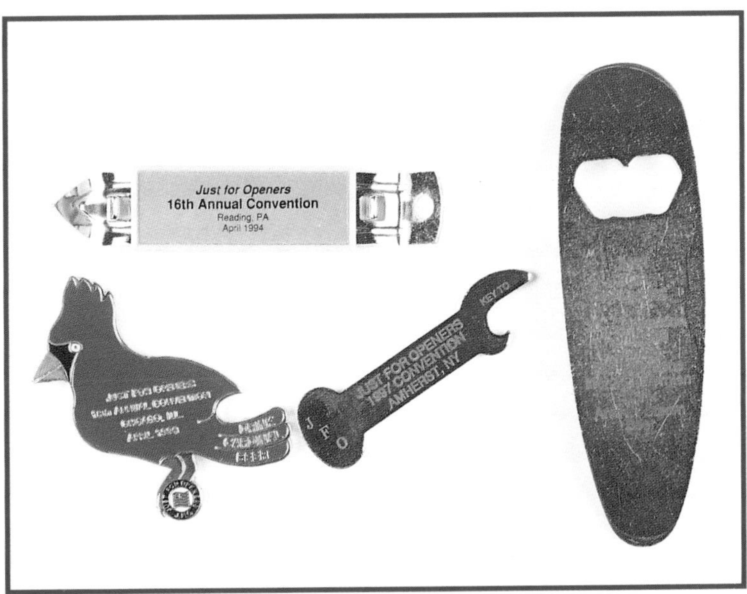

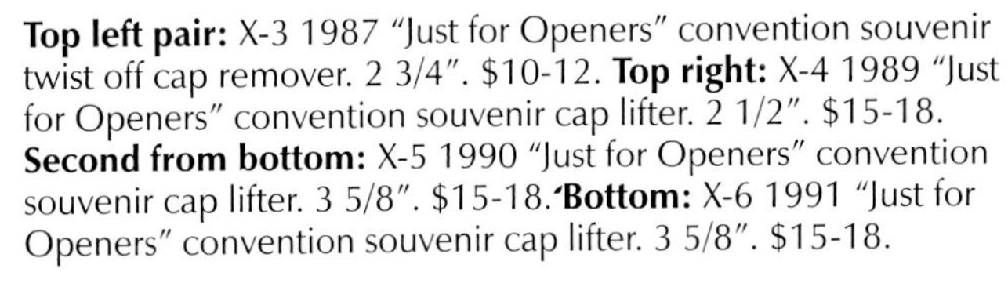

Top left pair: X-3 1987 "Just for Openers" convention souvenir twist off cap remover. 2 3/4". $10-12. **Top right:** X-4 1989 "Just for Openers" convention souvenir cap lifter. 2 1/2". $15-18. **Second from bottom:** X-5 1990 "Just for Openers" convention souvenir cap lifter. 3 5/8". $15-18. **Bottom:** X-6 1991 "Just for Openers" convention souvenir cap lifter. 3 5/8". $15-18.

Top left: X-7 1994 "Just for Openers" convention souvenir Cap lifter/can piercer. 4". $5-8. **Bottom left:** X-8 1996 "Just for Openers" convention souvenir cap lifter. 2 7/8". $15-18. **Middle:** X-9 1997 "Just for Openers" convention souvenir cap lifter. 3". $3-5. **Right:** X-10 1998 "Just for Openers" convention souvenir cap lifter. 5". $3-5.

In a Class by Itself

It doesn't look much like a bottle opener—but it is!

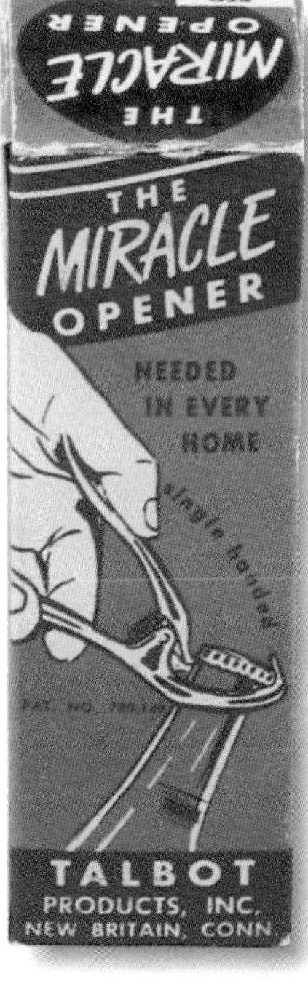

Z-1 "Miracle Opener." Pliers style cap lifter holds cap in jaws when removed. American patent 2,651,511 issued to Joseph A. Talbot, May 1, 1951. Manufactured by Talbot Products, Inc., New Britain, Connecticut. The packaging for the opener is marked PATENT NO. 780,149. That is the patent application number not the patent. 4 3/4". $30-40.

Part 2

Other Advertising Openers

Opener Manufacturers' Openers

An opener manufacturing company's best way to advertise was to use their own products featuring their company name. The manufacturers would often assign catalog numbers to their openers as well as catchy slogans. Besides opener manufacturing companies, advertising sales specialty firms also used this method.

Top row: B-1 Enameled opener by Electro-Chemical Engraving Co. Manufactured before 1920. After Prohibition (1933), the company produced types M-3 and M-73 (slide-out openers). $20-25; B-6 made by G. P. Coates & Co. (also made type B-46). $8-10; B-8 Cigar Box Opener by L. F. Grammes (also made type B-39). Example shown is numbered on the back and has a "Finder Rewarded" slogan. $10-12. **Second row:** B-31 "Crown Opener" made by Williamson Mfg. Co. (No. 110). $10-12. B-29 "Crown Opener" by Williamson (No. 108). $10-12. **Third Row:** B-14 "Improved Crown & Seal Co." $3-5; B-21 "The Presto Bottle Opener and Gas Tank Key." $8-10; B-27 Enameled opener by J. K. Aldrich. $20-25. **Fourth Row:** B-24 Vaughan Novelty Mfg. Co.'s "Item #3." $5-8; B-2 Enameled opener by J. K. Aldrich with ad for "Advertising Metal Material." $20-25. **Bottom:** B-2 Enameled opener by J. K. Aldrich with ad for "Bottle Openers." $20-25.

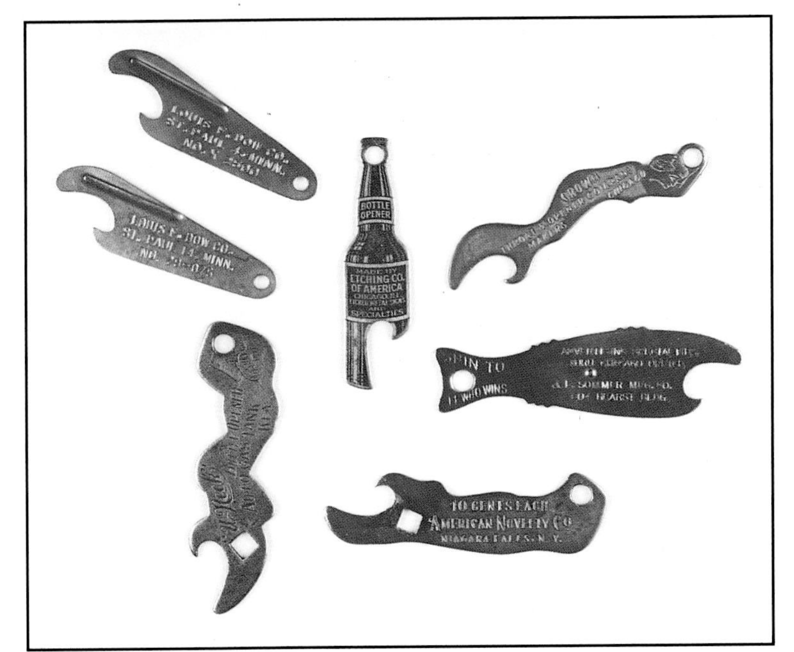

Top left: Two different A-23 Dow pin openers by Louis F. Dow Co. Marked ST PAUL 4 MN NO. X 2500 and ST. PAUL 14 MN NO. 29-076. $8-10. **Middle:** A-24 enameled bottle by America Etching Co. $15-20. **Right, top to bottom:** A-1 clothed lady advertising "Crown Throat & Opener Company Makers Chicago." $15-20; A-20 fish from John L. Sommer's Chicago office. Sommer was headquartered in Newark, New Jersey. $10-12; A-30 dancer legs opener could be purchased through the "American Novelty Co. Niagara Falls, N. Y. 10 Cents Each." $15-20. **Bottom Left:** A-4 lady "U-Neek" Bottle Opener & Auto Gas Tank Key. Also by the Crown Throat & Opener Company. $15-20.

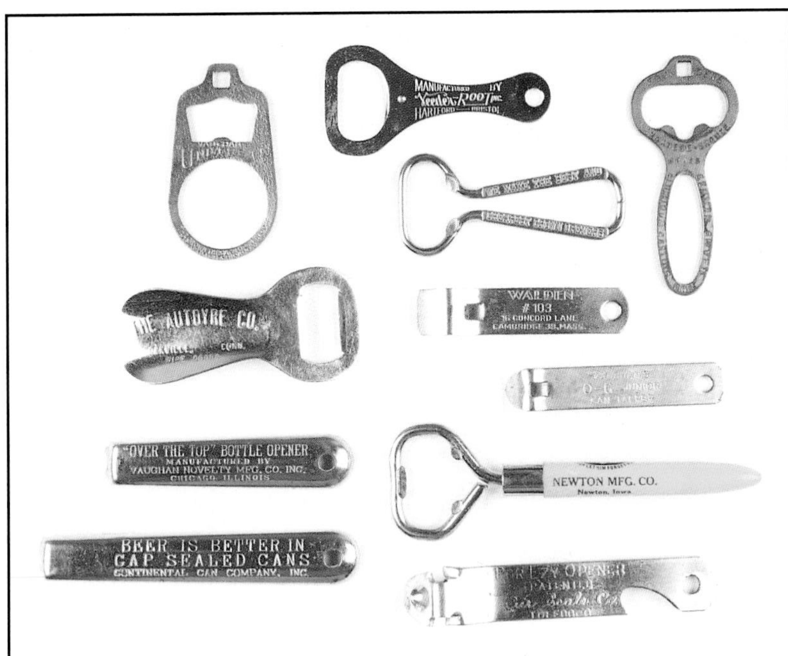

Left, top to bottom: Vaughan's "Universal" opener with gas key made by Crown Throat & Opener Co. $20-25; N-11 shoehorn opener by Autoyre (pronounced "Auto Wire"). $8-10; H-2 "Over The Top" bottle opener by Vaughan. $3-5; H-3 "Over The Top" advertising "Continental Can Company" for "Cap Sealed Cans." $3-5. **Second column, top to bottom:** C-10 spinner opener by Veeder-Root Inc. of Hartford-Bristol (Connecticut). $3-5; E-8 wire opener made by Autoyre. $3-5; G-6 cap lifter by Walden of Cambridge, Massachusetts. $1-2; J-11 called "Vaughan's O-G Junior Can Tapper." $1-2; M-51 by Newton Mfg. Co. $3-5; I-24 heavy steel can piercer called "Por-Ezy Opener" by Air Scale Co. of Toledo, Ohio. $12-15. **Top right:** "Quick & Easy" made by Erie Specialty Co. of Erie, Pennsylvania, with gas key. Back side marked FOUNDERS OF MANGANESE-BRONZE BRASS GERMAN SILVER AND ALUMINUM CASTING. $15-20.

Beautifully engraved opener produced by the Crown Cork and Seal Co. of Baltimore, Maryland. Produced as a convention souvenir from 1908. $75-100.

World's Fair, Knives, Wall Mounts, & Unusual Large Openers

All "World's Fair" openers shown are from the 1933-34 Century of Progress in Chicago. This fair certainly produced a greater variety of openers and corkscrews than any other World's Fair.

More N-4 knives are shown because they are very colorful and cross several collecting areas—beer, soda, and souvenir.

Wall mounts are popular with many collectors and some more unusual types are pictured. The wall mount with a corkscrew advertises White Rock Ginger Ale. The White Rock Mineral Springs Company started producing the drink in the 1890s. Another wall mount advertises Whistle soda. Whistle was invented in 1916 by Silvester Jones and was a very popular Midwestern soda. The wall mount openers shown are from the 1920s-1950s.

A selection of cast iron and kitchen openers round out this section. Kitchen utensils came in many forms and with a variety of attachments. The opener was the most common attachment and kitchen items saw their greatest popularity in the 1920s-1950s. Two of the can openers pictured are very unusual because they advertise beer. Some collectors would assume they should be in the first section of this book, but they are tin can openers not bottle openers.

World's Fair Openers. **Left side, top to bottom:** N-4 knife. $25-30; B type cap lifter. $12-15; Heavy cast iron opener with two heads. $50-60. **Right side, top to bottom:** M-73 slide-out opener and corkscrew. $75-100; Similar to an M-3 slide-out opener, but this one is twice as wide and only half as thick (M-3 examples do exist for both years 1933 and 1934: their value $20-25). $100-125; Combination jar opener and cap lifter by Kerr Glass Co. $12-15; Woman style opener with the original box. $35-40 (without the box $15-20). **Bottom middle:** Very unusual small whale marked STERLING. $75-100.

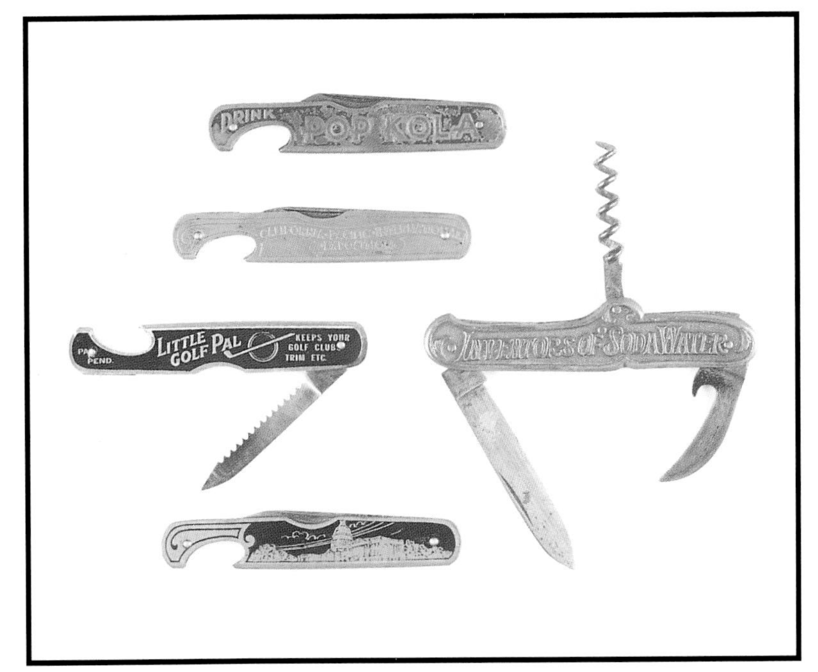

Left side, top to bottom: Type N-4 knives—"Drink Pop Kola" with "Drink Tick Tock" from the Atlanta Bottling Works on the reverse side. $35-40; 1935 California Pacific International Exposition in San Diego. $40-50; "Little Golf Pal" with "Designed and Distributed by Sam Sharrow Golf Pro Malden, Mass." on the reverse side. $40-50; Souvenir of Washington D. C. $35-40. **Right:** Knife advertising "Inventor of Soda Water" with "A & R Thwaites & Co., L^td^" on the reverse. c.1900. $125-150.

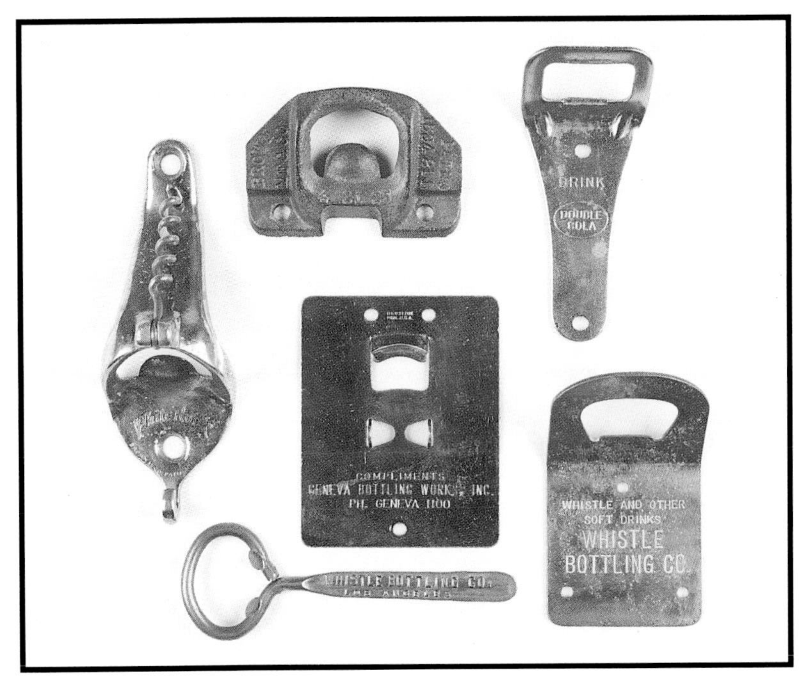

Left: White Rock Ginger Ale opener and corkscrew by Brown Mfg. Co. of Newport News, Virginia. $60-75. **Middle, top to bottom:** One of Brown Mfg. Co's first openers. Marked PAT NO 1534211 4 21 25 BROWN MFG CO. $60-75; B & B, St. Paul wall mount for the Geneva Bottling Works of Geneva, New York. $40-50; E-4 Wire opener from the Whistle Bottling Co. of Los Angeles. $12-15. **Right:** A flat C-type opener bent 45 degrees with two reinforcing ribs added for Double Cola. $35-40; O-19 type Whistle wall mount. $40-50.

Left: Cast iron Hebe Ginger Ale (Providence, Rhode Island). A c.1910-1930 opener with a bottle stopper and a crown size button on top. $75-100. **Middle pair:** Cast iron crown opener with an actual bottle cap attached. $35-40; "C&C Super" heavy brass opener with cone top can. Supposedly from Cantrell & Cochrane of New York. c.1940-1950. $75-100. **Right:** Fold-up wire type for Pureoxia Ginger Ale marked PAT APLD FOR. Pureoxia was a ginger ale made by the Moxie Co. of Boston, Massachusetts. $60-75.

Left three: Potato masher with a one digit phone number. $15-20; French whisk advertising a grocery store. $12-15; Two tablespoon bowl which is much larger than a standard type F-2 bowl and with a one digit phone number. $15-20. **Right side, top to bottom:** Can opener advertising Gutsch Brewing Co. of Sheboygan, Wisconsin. $40-50; "Pabst Malt Syrup" can opener. $40-50; Heavy cast iron ice shaver. Somewhat unusual with bottle opener. $15-20.

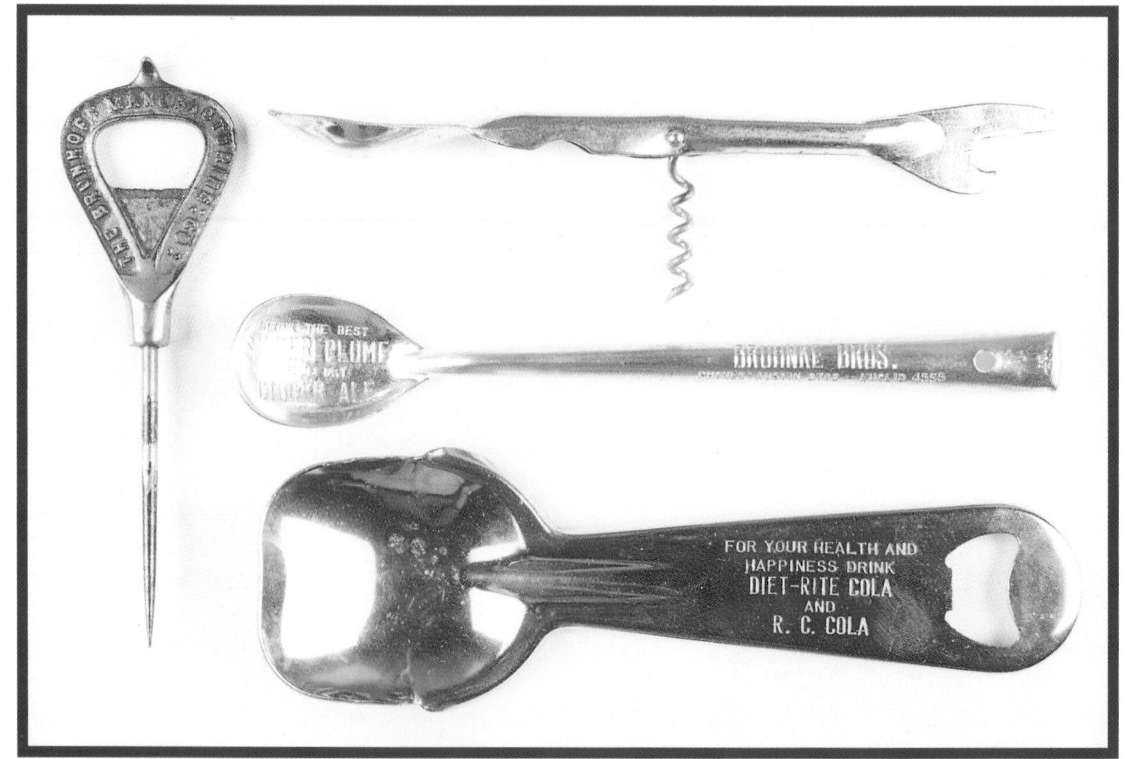

Left: Early 1900's ice pick and loop-seal remover, like type F-14, by "The Brunhoff Manufacturing Co." of Cincinnati, Ohio. Brunhoff was a large manufacturer of counter top cigar cutters and many different advertising give-aways. $30-40. **Right, top to bottom:** Ice shaver and corkscrew from the "National Nu Grape Company." Very unusual design. $60-75; Silver-plated type F-29 for "Silver Plume Pale Dry Ginger Ale." $12-15; A fairly common opener and ice cream scoop type. This one is very uncommon because of soda advertising for "Diet-Rite Cola" and "R. C. Cola." $20-25.

Unusual Flat Openers

This category serves as a catch-all for the many numerous odd types of flat openers. The collector should be aware that many odd shapes exist without beer or soda advertising and they are desirable. Some of the types have been shown in the first section of the book but are included here because they are general soda items.

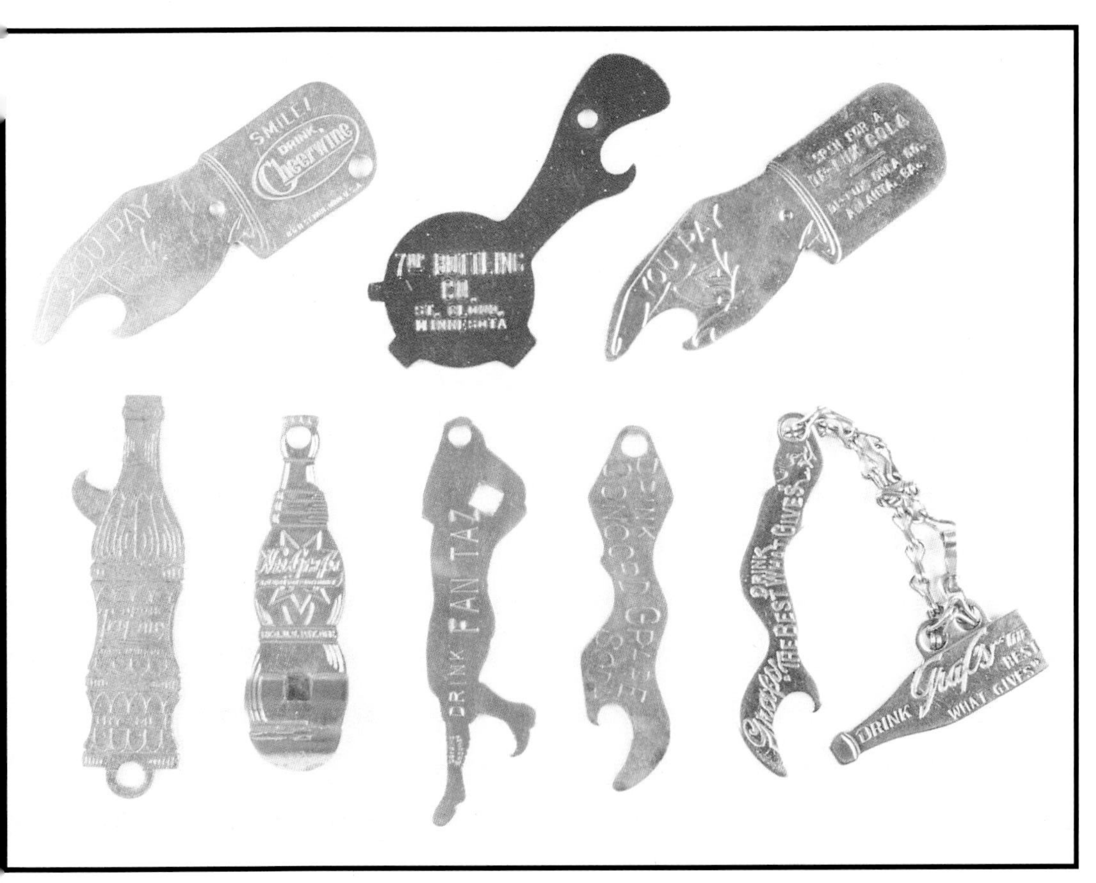

Top, left to right: A-21 finger spinner for "Cheerwine." $20-25; A-39 turtle opener. "7-Up." $20-25; A-21 finger spinner "De-Lux Cola, Atlanta, Ga." $15-20. **Bottom, left to right:** Bottle shaped "Try-Me" beverages. Fairly common with Try-Me on one side. $12-15 (on two-sides $20-25); Nu-Grape bottle shaped opener with a curved cap lifter. $12-15; Type A-9 baseball player advertising "Fan Taz." The unusual thing about this opener is ADV. NOV. CO. CHICAGO is marked on the leg. It is the only A-9 known with a maker's name. $50-60; A cross between an A-1 and an A-5 figural lady opener for "Concord Grape Soda." This is the only advertising seen on this type. $15-20; A-1 "Graf's Soda" with a small bottle attached with a key-ring. Sometimes companies would give two items away instead of one. $30-35.

Left to right: Similar to type E-30, a souvenir of the "Lighthouse and Cliffs Rock Island, R. I." $10-12; Next three openers are customized pieces with pearl handles added. The last two are former beer advertising openers that now have their ads covered up. Each one looks different. $20-25; A multi-tool opener, button hook, gas key, 2-inch ruler, screwdriver tip, and key ring opener designed and made by "E. Roberts" of St. Louis, Missouri. $15-20; Opener, button hook, and key ring very similar to type B-72 with advertising for "Andrew V. Serio Advertising Novelties 139 N 4th St. Phil Pa." $15-20.

***Top:* Top, left & right:** Goodyear Airplane. $60-75; Holsum Bread. $20-25. **Bottom, left to right:** A-B Stove Co. of Battle Creek, Michigan. Opener, gas key, and screwdriver tip. Looks somewhat like a pot-bellied stove. $20-25; Type A-12 made of "Duralumin" used in construction of the airship "Akron." $20-25; "Arkansas Bottle Fish Bites Best on Full Bottles" from Lincoln Prod. Co. of Lincoln Park, Michigan. Unusual fish figural only seen with this ad. $35-40; Chicken figural with screwdriver tip at top of head. Always seen with chicken ads. $10-12; Two cigar pipe cleaners. A very unusual combination tool with the pipe cleaner easily bent or missing. Both would actually be considered "pipe" figurals or "A" types. One has an ad for "Moonshine Smoking Tobacco" of Winston-Salem, North Carolina. $50-60; The other has a gas key and directions "For Pull Out Pipe Cleaner." $35-40.

***Center:* Top, left to right:** "The Scruit" with opener and four screwdriver tips. Marked FOR CIGARETTE LIGHTERS LOOSE SCREWS. $10-12; Type B-9 with unusual ad for "Corbin" and a big screw pictured. $20-25; Independent Lock Co. of Fitchburg, Massachusetts, came up with the novel idea of adding a bottle opener to a standard key. Back and front shown. $10-12. **Bottom, left to right:** Combination opener, gas key, screwdriver tip, and small jar opener made by the "E. O. B. Mfg. Co. Grand Rapids, Mich." $20-25; Combination opener, gas key, screwdriver tip, and small jar opener. Marked SYRACUSE AUTO SUPPLY CO. $20-25; Bolt wrench with opener and two size gas keys for the "Meteor Gas Co. Middle Haddam, Ct." on one side and "Meteor Auto Tank Co. New York" on the other side. $20-25; Combination opener, button hook, gas key, and key ring made by the "Wakefield Mfg. Co. of Wakefield, Neb." $20-25; A multi-tool with opener, button hook, screwdriver tip, gas key, key ring, 2-inch ruler, and knife sharpener with ad for "Weaver's of Ligonier, Pa." $15-20; "B" type for "Big Diamond Milling Co." of Minneapolis, Minnesota. Tools include an opener, screwdriver tip, gas key, and a large gas key wrench. Made by Quimby Mfg. Co. of Minneapolis. $35-40.

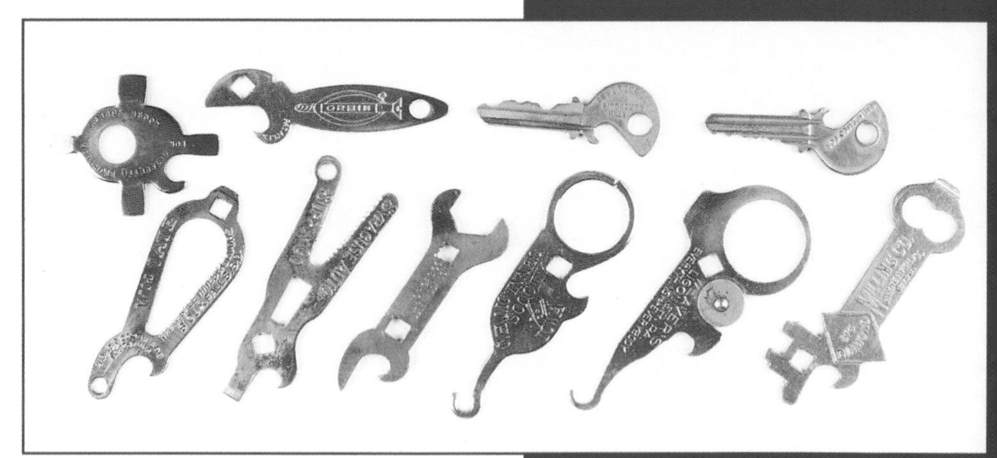

***Bottom:* Left:** Coffee can opener for "George Washington" coffee showing four different size cans. $30-35.
Middle, top to bottom: "Manitou" Ginger Ale opener with an unusual extension. $35-40; Another coffee can opener very similar to type B-21 with ad for "Allentown Transfer Co." $15-20; "Vaughan's Perfect Scissors Sharpener" were a neat idea for a "C" type opener with no opener. $15-20; Vaughan's "Dainty Milk Bottle Opener." $5-8. **Right, top to bottom:** Similar to type M-8 but smaller with an ad from "Platterville Bottling Works." $10-12; Three "B" type openers with screwdriver tips. Ads are from a "Billiard" place in Lewistown, Mt., "Carney & Co" funeral directors, and "Country Club Pale Ginger Ale" of Newton, Mass. $12-15.

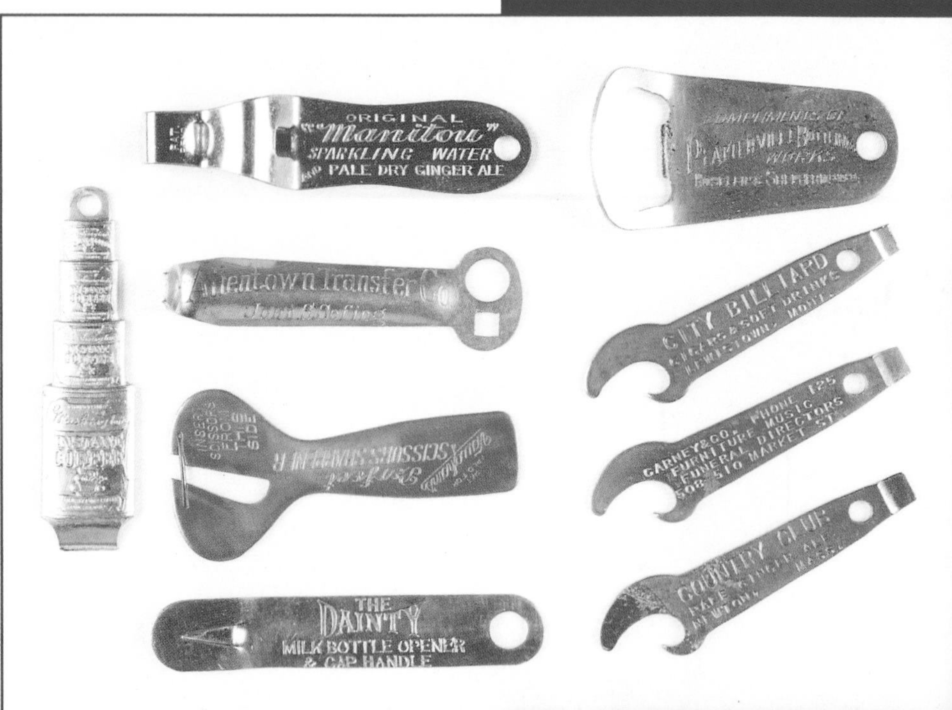

Coca-Cola Openers

Coca-Cola was invented in 1886 in Atlanta, Georgia. Coca-Cola is considered the king of all soda-pop collectibles and it is certainly true with openers. Not only did Coca-Cola make by far the widest variety of openers, the prices that these openers bring are the highest or near the highest for any brand whether it be soda or beer. Over 500 different Coca-Cola openers exist counting variations. Most are pre-1970 and like Anheuser-Busch today, Coca-Cola keeps adding many new types every year.

***Left:* Top to bottom:** B-13 type with corkscrew. $60-75; B-18 type New Castle, Pennsylvania. Bottling Company. $40-50; B-19 and B-22 type, both from the My-Coca Company of Birmingham, Alabama. The owner Diva Brown claimed My-Coca was made from the original Coca-Cola formula (on the back of both openers). Coca-Cola eventually put her out of business. c.1900s-1910s. $75-100; B-22 type from Philadelphia. $40-50; B-23 type from Weldon, North Carolina. $50-60; B-24 "Never-Slip" from Hopkinsville, Kentucky. $35-40.

***Below:* Left:** A-1 type clothed lady figural. $175-200. **Middle, left top to bottom:** A-17 Lion head figural with Coca-Cola in block letters. $100-125; A-43 Horse Head figural from Waverly, Iowa. $200-225; A-23 Dow pin. $35-40. **Middle:** Two different fish spinners. One with "Sterilized Bottles" and one without. $125-150. **Right:** A-21 brass spinner is a recent "fantasy" item. Not authorized by Coca-Cola. Unlike original 1940s-1950s spinners, this one has the "spinner knob" punched off center, no B & B markings (the original maker), and a hand line appears at the top middle of the opener. Original spinners did not have this hand line. $3-5; A-53 Dancer legs. $150-175. **Bottom:** C-12 type with embossed baseball and bat plus slogan "Drink Of The Fans." $100-125.

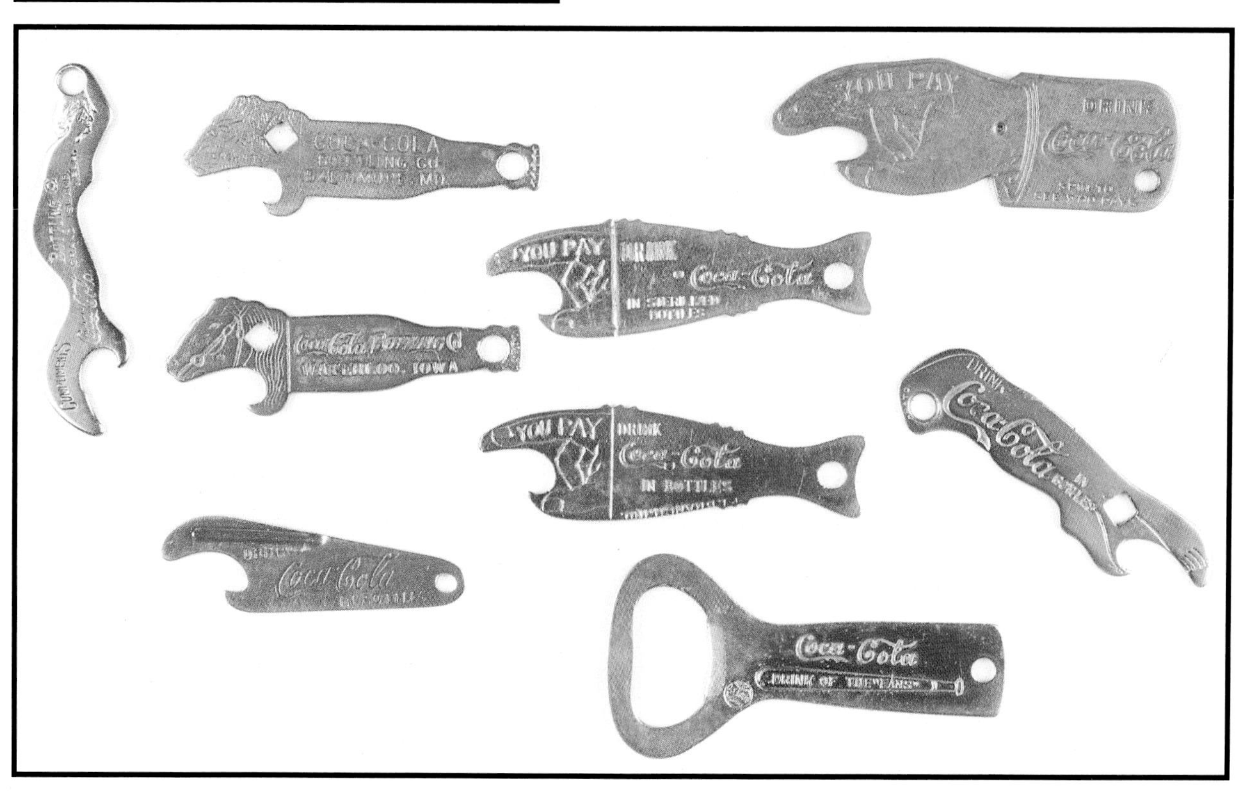

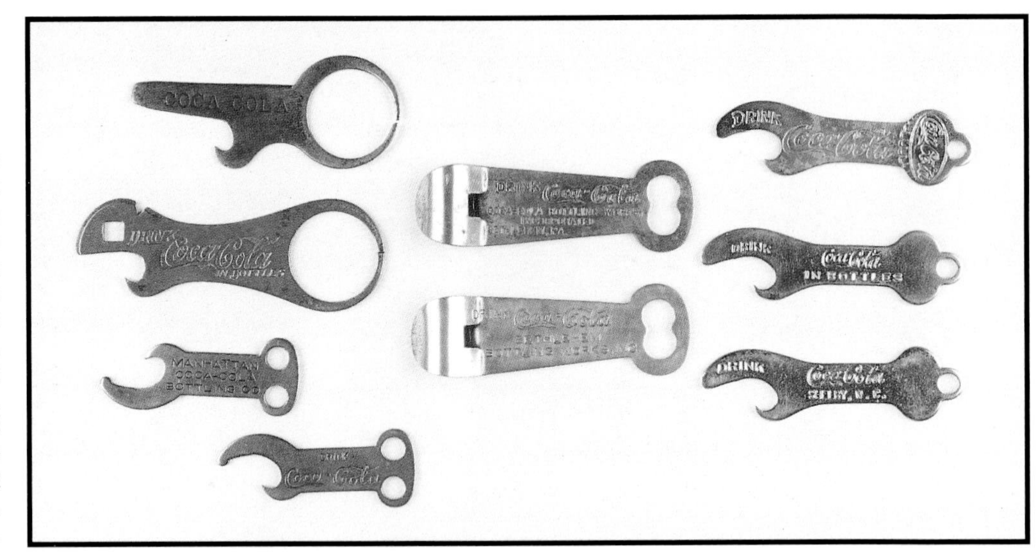

Left, top to bottom: B-6 type with just Coca-Cola in block letters. $75-100; B-42 type with cigar box opener. $50-60; Two B-32 types: script and block Coca-Cola. $75-100. **Middle:** Two different B-30 types with formed cap lifters. $50-60. **Right, top to bottom:** Most common Coca-Cola "B" type. Tougher to find with city and town names. Top B-14 is the most commonly seen B-14 but the back side has Easton, Pennsylvania. $20-30.

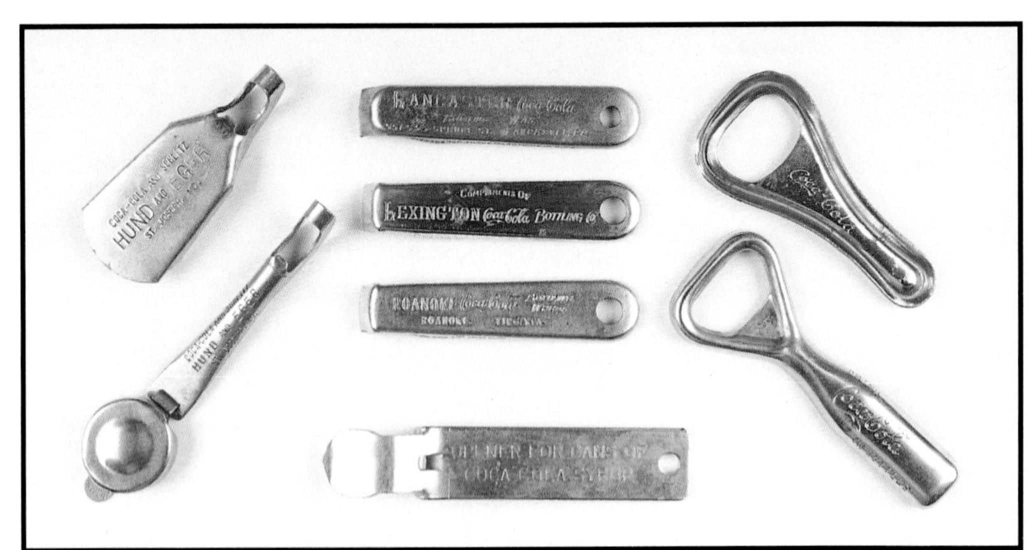

Left: N-62 with both Coca-Cola and Schlitz from the Hund and Eger Bottling Co. of St. Joseph, Missouri. Worth more as a Coca-Cola collectible than Schlitz. $50-60; G-25 opener and cap resealer. Advertising is like the N-62. $75-100. **Middle, top to bottom:** Three different H-2 types from Lancaster, Pennsylvania; Lexington; and Roanoke, Virginia. $40-50; Coca-Cola syrup can opener. $30-40. **Right:** G-13 "Vaughan Perfection" opener. Some years back a large pile of these turned up but has long since dispersed. $20-25; G-1 type with the generic slogan "Drink Coca-Cola in Bottles." $50-60.

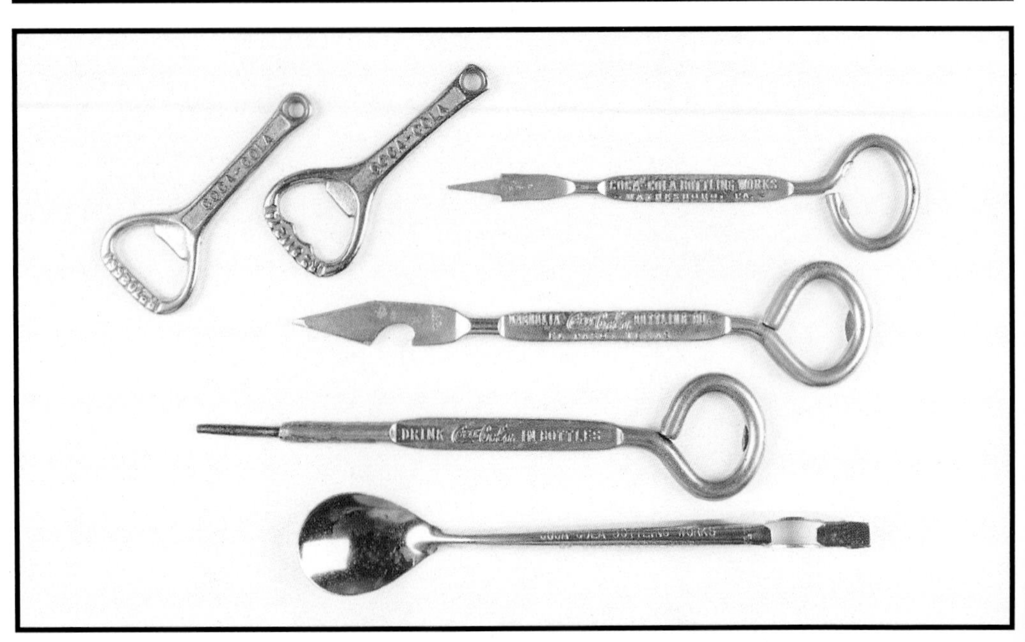

Left top: Two cast iron openers with the English patent number and slightly different shapes. Cast iron openers of this type date from 1900 and through 1950. $50-60. **Right, top to bottom:** F-6 type 1912 patented 4-in-1 opener from Waynesboro, Pennsylvania. $75-100; F-5 heavy steel opener from the Magnolia Coca-Cola Bottling Co. of El Paso, Texas. $75-100; F-27 type two-piece constructed opener and ice pick from Baltimore, Maryland. $60-75; F-15 type from Greenwood, Mississippi. $60-75.

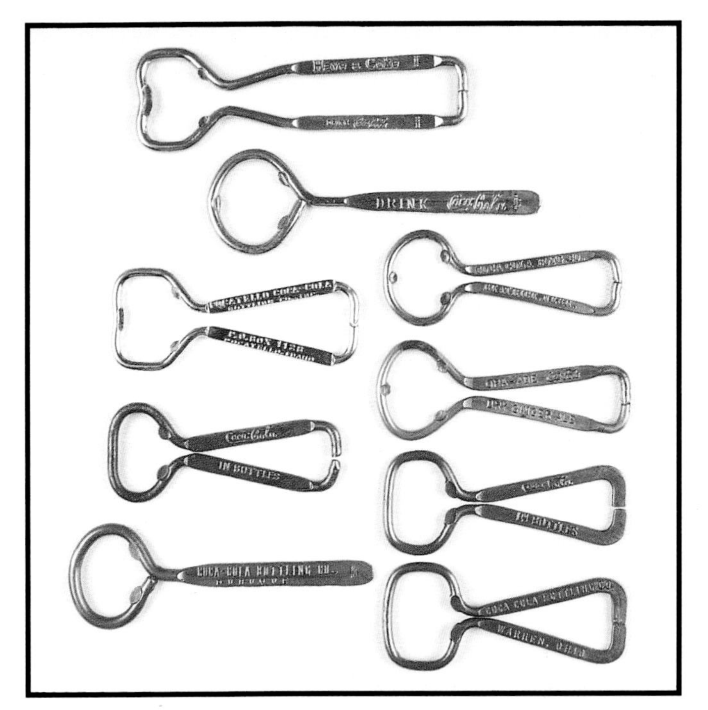

Top: E-11 type. $30-35; E-4 type (single strand) with "Sterilized Bottles" slogan. $60-75 **Left, top to bottom:** E-14 type from Pocatello, Idaho. $20-25; E-8 type with generic Coca-Cola "Delicious and Refreshing" slogan. $35-40; E-4 type from Dubuque, Iowa. $40-50; **Right, top to bottom:** E-9 type with rounded top (oldest E-9 type) from Beatrice, Nebraska. $35-40; Newer E-9 and this one probably Canadian (Paschal Beverages). $20-25; E-7 type with generic Coca-Cola slogan. $15-20; E-7 type from Warren, Ohio. $35-40; Note that many newer E-14 types are not shown and most are only worth $1-2.

Left top: O-4 Vaughan "Never-Chip" wall mount. $100-125. **Left bottom:** O-19 wall mount from San Bernardino, California. $75-100. **Right top:** O-2 Erickson enameled wall mount. $40-50. **Right bottom:** "Hoof" cast iron wall mount. $100-125.

Top left: N-43 type knife. $50-60. **Bottom left:** N-29 "Bullet Pencil" from Starr Brothers of Akron, Ohio. $100-125. **Middle:** Coca-Cola bottle shape from Germany. $30-35. **Right.** Schrade corkscrew knife with small Coca-Cola bottle attached. $75-100.

Pepsi-Cola Openers

Pepsi-Cola was invented in 1898 in New Bern, North Carolina. Due to instability in the early years, few pre-1920 Pepsi openers exist, and most are from the 1930s onward. Unlike Coca-Cola with its wide array of types, Pepsi is mainly confined to "B" key shaped, "E" wire openers, and wall mounts. Type "A" figurals are known to exist and are very rare except for the bottled shaped "A-29." Counting name variations, one could probably collect about 100 different pre-1970 Pepsi openers. Collectors of course usually prefer double dot (=) Pepsi items to single dot (-).

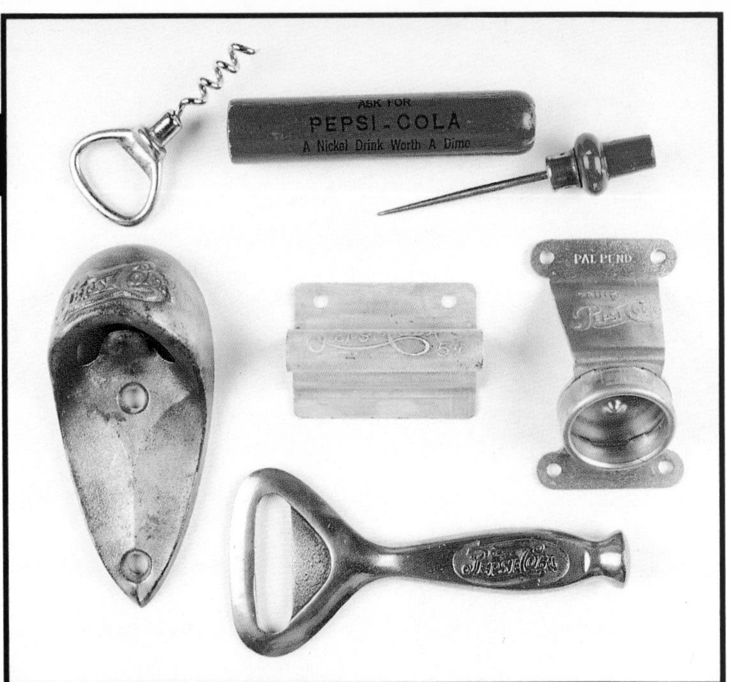

***Above:* Top:** Rare combination ice pick, corkscrew, and opener. $100-125. **Middle, left to right:** "Tear-drop" cast iron wall mount. $35-40; O-4 "Never-Chip" type. $60-75; Round wall mount cap lifter. Only seen with Pepsi-Cola. $60-75. **Bottom:** M-19 heavy brass opener. $40-50. Beware: This opener has been reproduced. On reproductions the muddler end is flat and not rounded.

***Left:* Left, top to bottom:** C-19 type. $12-15; G-1 type with Pepsi=Cola script. $50-60; G-1 type from Muncie, Indiana. $40-50. **Right, top to bottom:** C-13 type. $15-20; C-34 type. $12-15; N-9 type with cap resealer. $30-35. **Bottom:** F-4 type from Princeton, West Virginia. $35-40; F-2 type with Pepsi-Cola in block letters. $35-40.

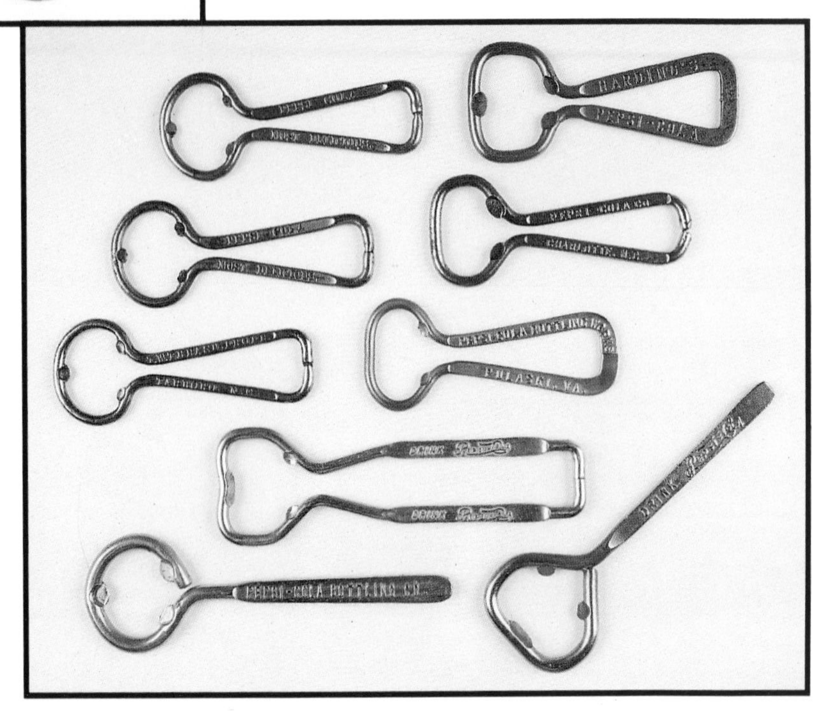

Left three: Top two are the same E-9 type opener. One has "Very Healthful" slogan on reverse. $35-40; E-9 shown is from Tarboro, North Carolina. $40-50. **Right three:** E-6 type. Harding's Pepsi-Cola. $15-20; E-8 type from Charlotte, North Carolina. $40-50; E-25 type from Pulaski, Virginia. $40-50. **Bottom three:** E-11 type. $10-12; E-4 type at bottom is very unusual in that "Arrow Beer" is advertised on the back side. $25-30; E-4 type. $15-20.

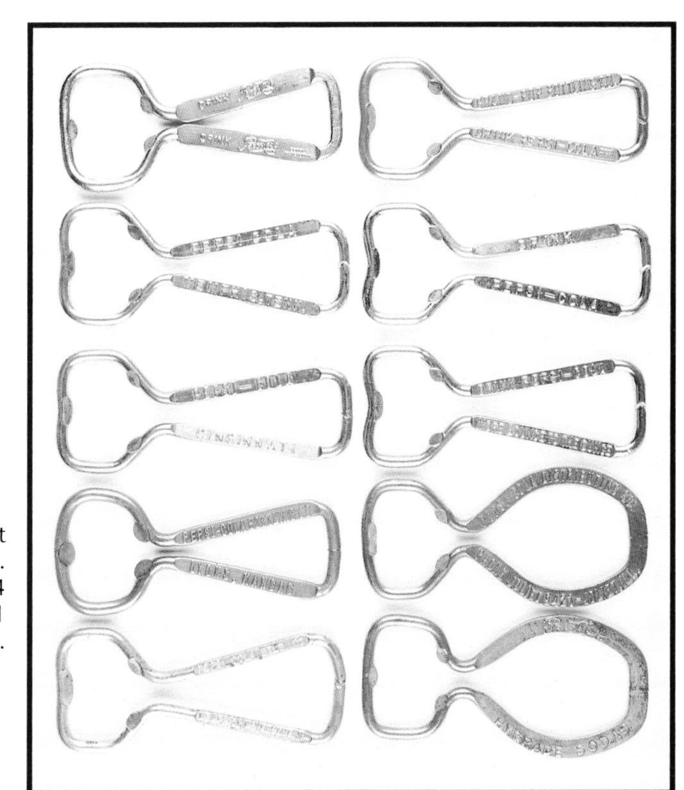

Left five: All E-14 type with slight variations in type or wording. $5-10. **Right five:** Top three are more E-14 variations. $5-10; Bottom two are E-1 types with different ads. $20-25.

Left: Two B-40 type openers that are pretty common. Difference is in slogan on the back. Top one is probably 1930s and the other 1940s. $12-15. **Middle:** B-18 type with Pepsi-Cola and Van Doren's Ginger Ale. $50-60; Bottom three openers are slight variations of each other and all from the Charlotte, North Carolina, area. $50-60. **Right:** Two B-14 types with like ads. Pepsi-Cola advertised on the reverse of each. $15-20.

Left: E-5. Pepsi bottle shown has a Seven-Up bottle on the reverse. Very unusual with two different bottles. $60-75. **Middle:** H-8 type lithographed bottle. $35-40 in excellent condition; L-1 Pepsi push-button can opener made in West Germany. $40-50. **Top:** E-5 type with a Pepsi-Cola bottle cap. $35-40. **Right, top to bottom:** I-27 folding cap lifter and can piercer. $12-15; H-5 over-the-top bottle opener. $20-25; Cast iron opener with the English patent number. $50-60; Wood sheath corkscrew and cap lifter from Charleston, South Carolina. $75-100.

Dr. Pepper & Nehi Openers

Dr. Pepper was invented in Waco, Texas, in 1895. Unlike Coca-Cola and Pepsi, very few pre-1920 openers exist (a 1890s folding legs corkscrew exists). The most highly collectible Dr. Pepper openers were produced in the 1930s-40s. Nehi was first produced in the 1920s and a nice selection of openers does exist including some very nice figurals. Like Dr. Pepper the best openers were produced in the 1930s-40s. A Dr. Pepper and Nehi collector could probably obtain about 50 different for each brand. In auctions, Dr. Pepper openers have been bringing very high prices.

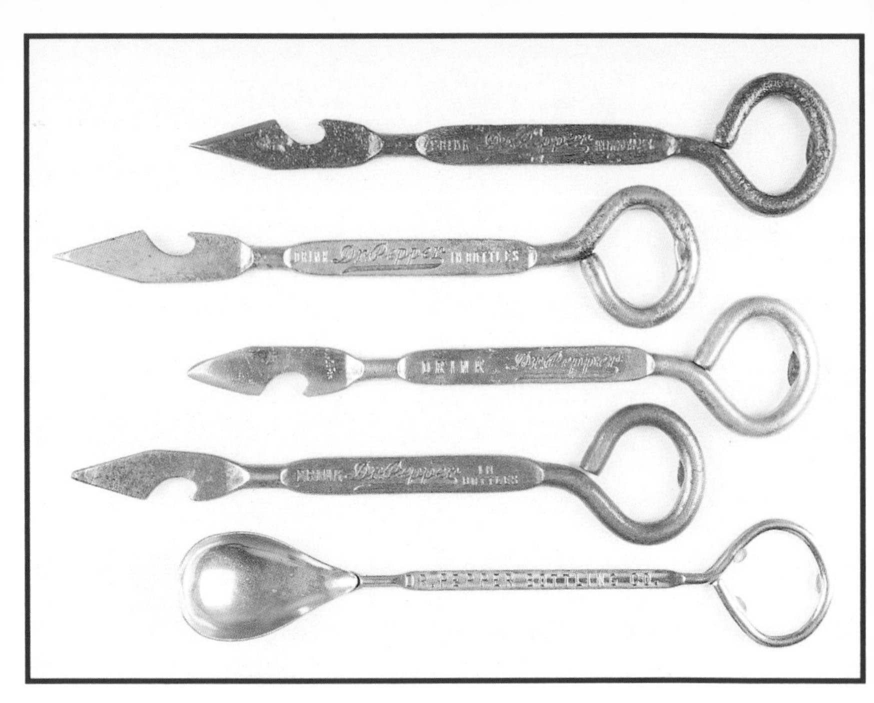

***Above:* Top four:** Four different Dr. Pepper advertising F-5 types: Top advertises Dr. Pepper with Nehi ad on the reverse. Second has generic Dr. Pepper slogan. Third is from Little Rock, Arkansas. Bottom is from Union Bottling Works. $75-100 in very good to excellent condition. **Bottom:** F-2 from Arkansas City Dr. Pepper Co. $75-100.

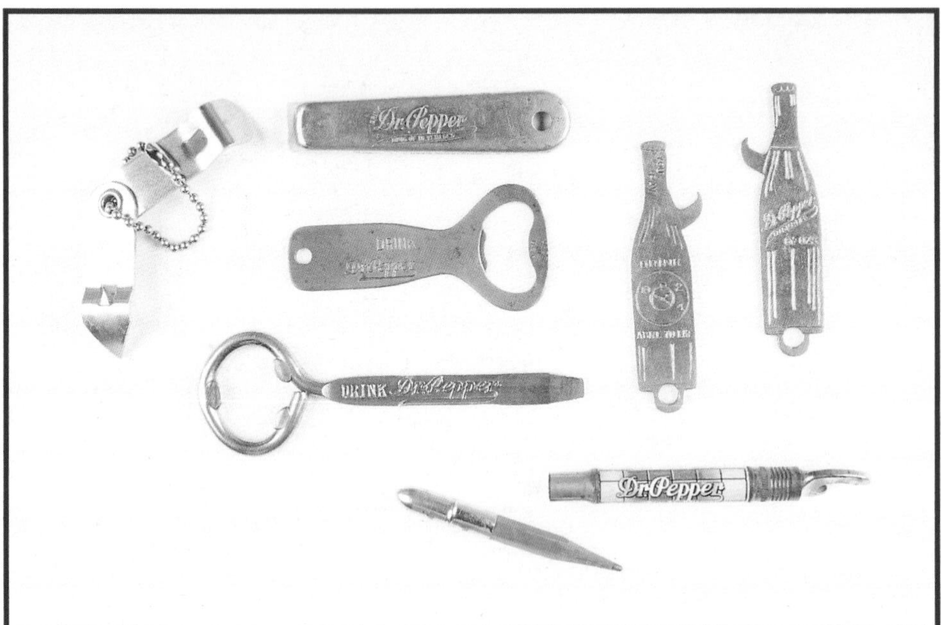

***Left:* Left:** I-27 folding cap lifter and can piercer. $15-20. **Middle, top to bottom:** H-2 over-the-top opener with "Good For Life" slogan. $50-60; C-13 type. $50-60; E-4 single wire with screwdriver tip. $60-75. **Right:** Two A-29 bottles. Left is a "You Pay" spinner. $60-75; Right is regular bottle without spinner. $50-60; N-29 "Bullet Pencil." $100-125.

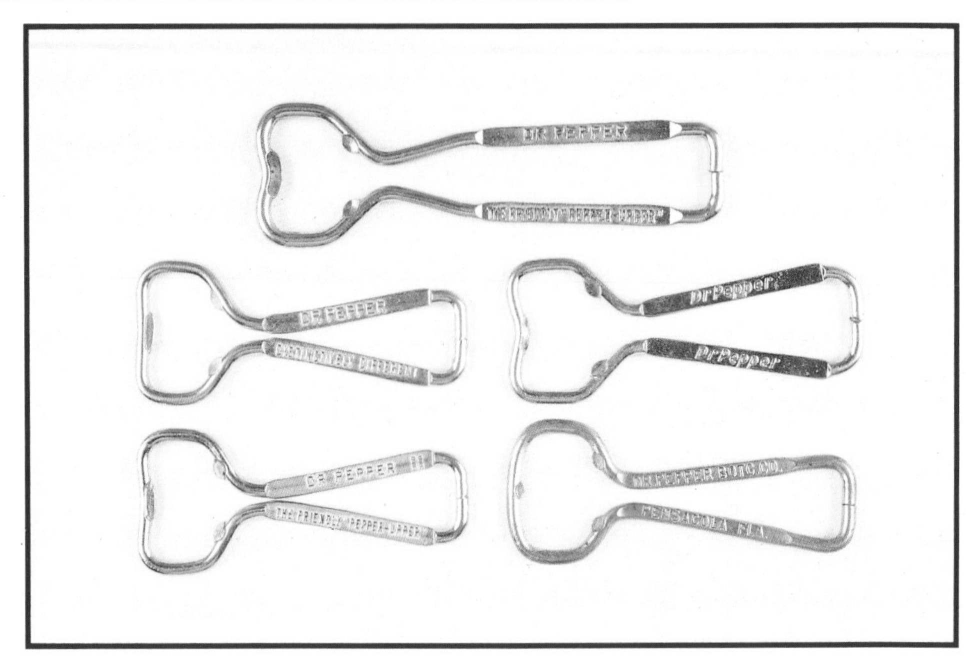

Top: E-11 long wire. $10-12. **Bottom:** Four E-14 types with various slogans. $8-10.

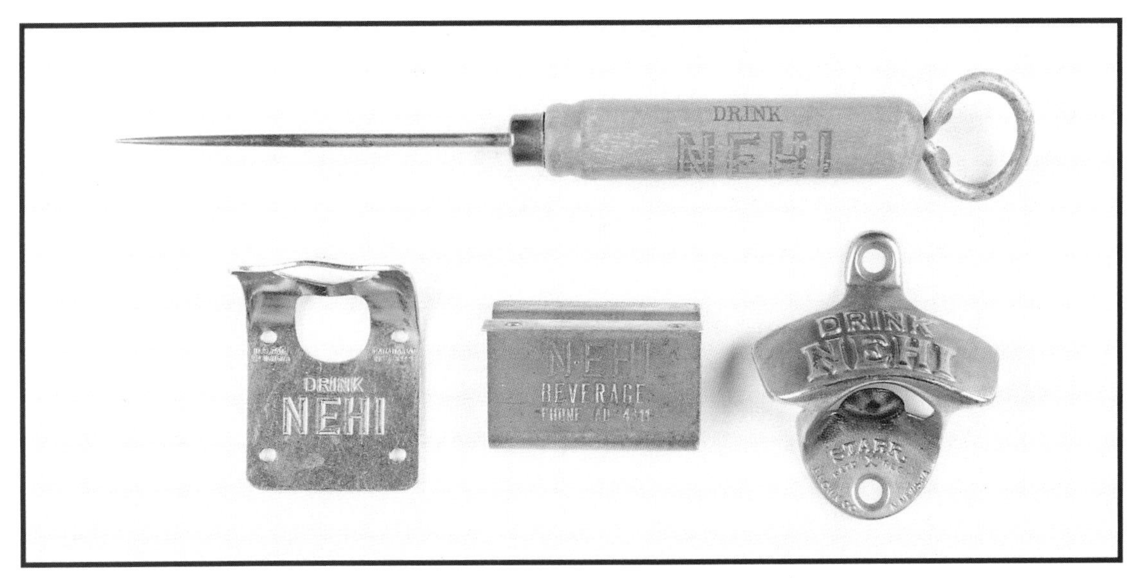

Top: F-24 type wood handle ice pick and wire opener. $40-50. **Bottom, left to right:** O-8 wall mount with double patent dates (US and Canada). $50-60; O-4 "Never Chip." $40-50; Starr X cast iron wall mount. Nehi is a fairly rare in this type. $60-75.

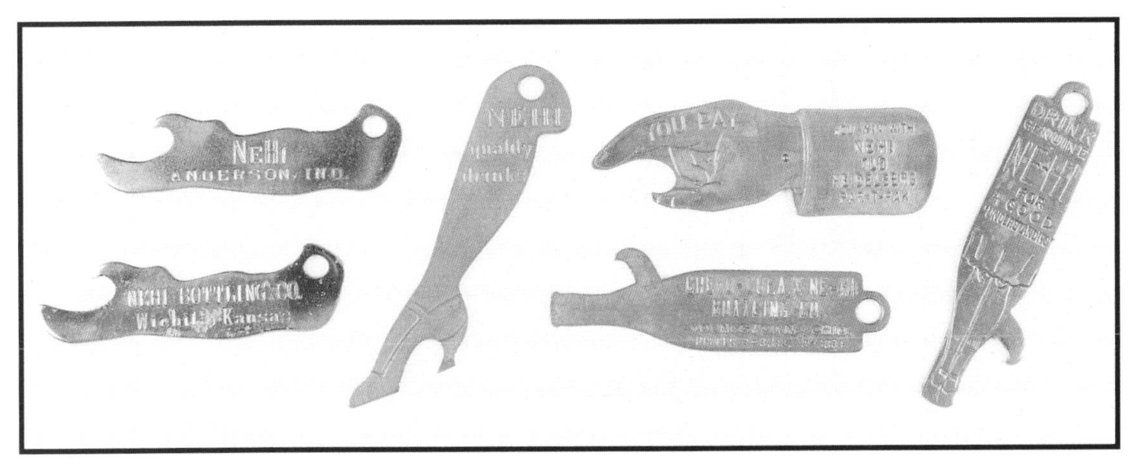

Left to right: Two A-30 Dancer Legs from Anderson, Indiana, and Wichita, Kansas. $40-50; Large leg similar to type A-54. $20-25; A-21 spinner with Par-T-Pak ad. $40-50; Two A-29. Figural legs and skirts shown on the front of each. $35-40.

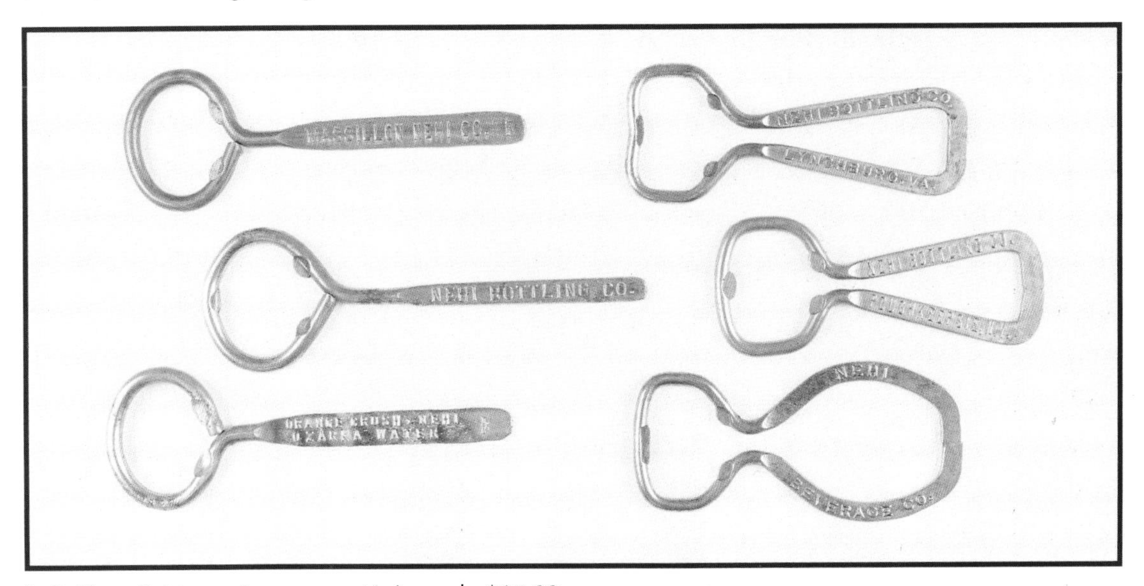

Left: Three E-4 type wire openers. Various ads. $15-20.
Right: Two E-6 type and one E-1 type. Various ads. $15-20.

Chero-Cola & General Soda Openers

Chero-Cola was developed in the early 1910s by the Union Bottling Works of Columbus, Georgia. Chero-Cola presents the collector with some nice figural openers and unusual pieces. There are about twenty different kinds. Hires Root Beer was first presented at the 1876 Centennial Exposition in Philadelphia. The most desirable Hires opener is a type M-2 lithographed; however, along with a handful of others, Hires had very few openers. Orange Crush was invented in 1916 in Chicago. Few early openers exist with most of the more desirable ones being from the 1930s-40s. Like Chero-Cola, a collector could probably find about 20 different types and name variations. Other brands shown are included to let collectors know that thousands of soda brands exist especially with ginger ale.

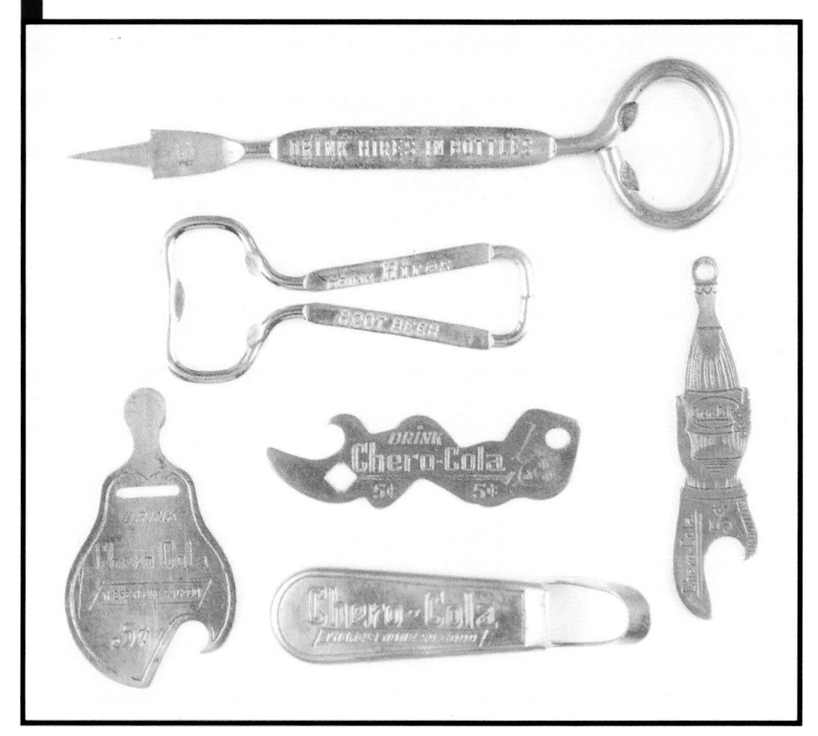

Top: F-6 Hires In Bottles opener and ice pick. $50-60; E-14 Hires wire opener. $5-8. **Bottom:** Chero-Cola watch fob and opener. $50-60; A-4 Chero-Cola lady figural. $40-50; G-5 Chero-Cola "Sealtite" cap lifter. $30-35; Figural hand holding a Chero-Cola bottle stamped 5 Cents. $40-50.

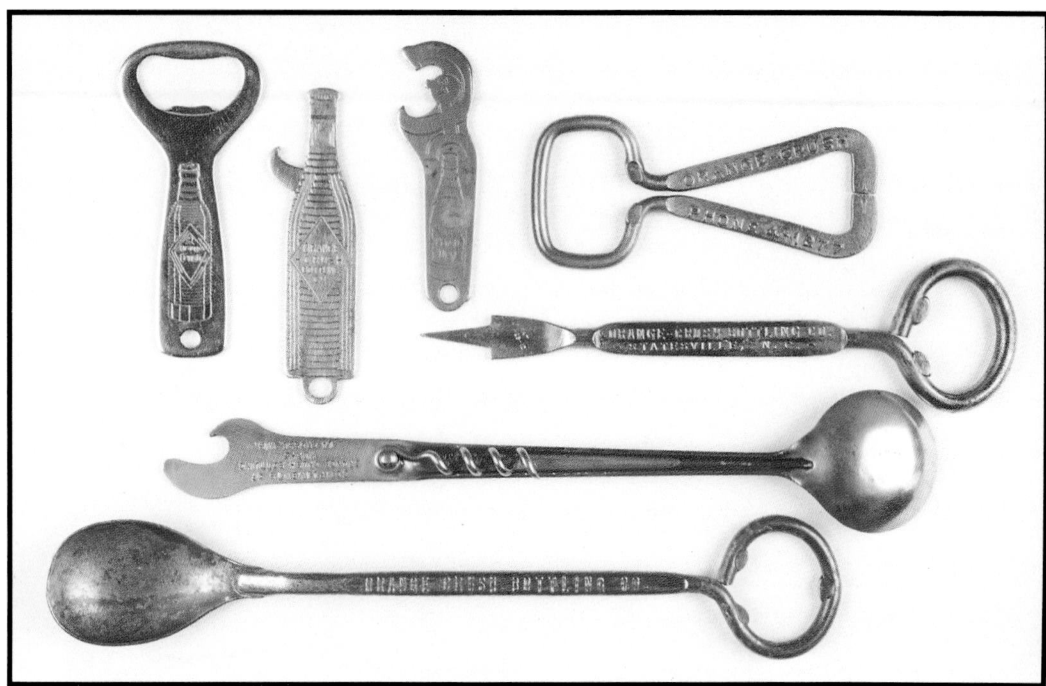

Top, left to right: C-13 Orange Crush. $12-15; A-29 Orange Crush. $12-15; "C Man" made exclusively for Orange Crush. $15-20; E-7 Orange Crush. $10-12. **Bottom, top to bottom:** F-6 1912 patent Orange Crush opener from Statesville, North Carolina. $20-25; F-1 opener with corkscrew from La Crosse, Wisconsin. $40-50; F-2 type spoon from Elizabethtown, Pennsylvania. $35-40.

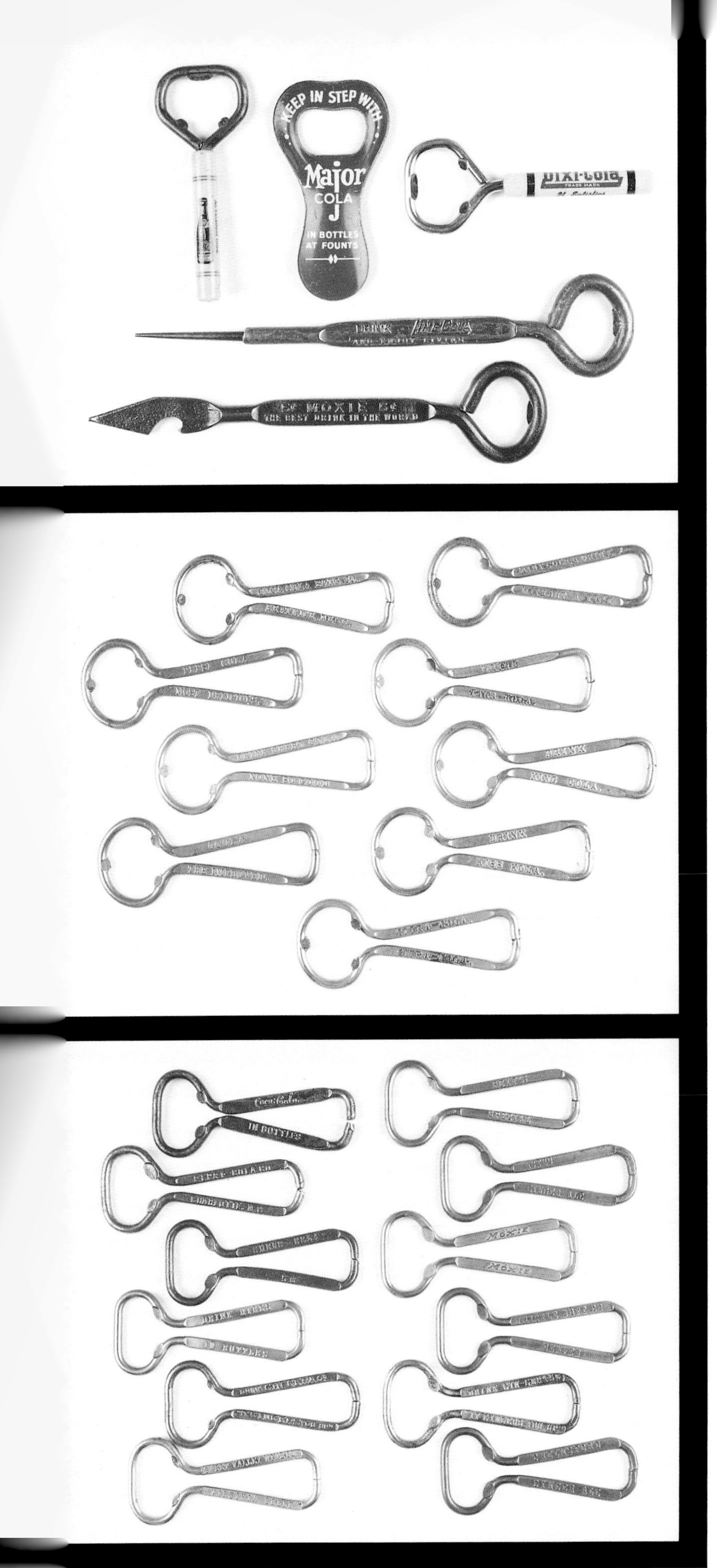

Top: E-5 type with 7-Up bottle from The Zip Co. of Chicago. $15-20; M-1 lithographed opener for Major-Cola. Hard to find these openers in nice condition. $40-50; E-5 type opener for Dixi-Cola. $12-15. **Bottom:** F-27 combination ice pick and opener for Lime-Cola. $20-25. F-5 type ice pick and double cap lifters for Moxie. $40-50.

A group of E-9 wire types from the 1910s and 1920s. Each advertises some kind of "Cola:" Coca-Cola from Beatrice, Nebraska; Pepsi-Cola; Chero-Cola from Greenville, South Carolina; Ocola from Western Ohio; Mint Cola; Taka-Cola; King Cola from Salem, Virginia; Kiss Kola from High Point, North Carolina; and Citra-Cola. E-9 was one of the earliest wire types and had its heyday from about 1915 to 1930. Values vary on these openers from $10-40 with the major brands being on the high end.

A group of E-8 wire types from the 1910s and 1920s. This type was used for a lot of general sodas, including: Coca-Cola; Pepsi-Cola from Charlotte, North Carolina; Chero-Cola; Hires; Gin-Cera (1-sided); Sweet Valley Wine Co. with Virginia Belle-Puritan Belle-Southern Belle; Reif's Special from Chattanooga, Tennessee; Nutro Ginger Ale; Moxie; Whistle from Denver, Colorado; Gin-Cera (2-sided); and Chelmsford Ginger Ale. E-8 was also one of the earliest wire types and had its heyday from about 1915 to 1930. Values vary on these openers from $10-40 with the major brands being on the high end.

Part 3

Information

American Patents

The story of the evolution of the advertising openers can easily be viewed by a look at the patents issued:

1876, February 1. P-1 and P-73. "Duplex Power Cork Screw." Patent 172,868 by William R. Clough.
1882, May 26. P-49 Waiter's Friend. German patent 20,815 by Karl Wienke.
1888, February 14. P-84 Bar mounted corkscrew. Patent 377,790 by Edwin Walker.
1891, February 24. P-3 Collapsing corkscrew. Patent 447,185 by Carl Hollweg.
1891, April 21. P-135 Bar mounted corkscrew. Patent 450,957 by Harry J. Williams.
1891, May 19. P-30 Bar mounted corkscrew. Patent 452,625 by Edwin Walker.
1891, July 14. P-2 The "Davis" corkscrew. Patent 455,826 by David W. Davis.
1893, July 25. P-8 Wood handle corkscrew. Patent 501,975 by Edwin Walker.
1894, February 6. D-6 & M-45. "Capped bottle opener." Patent 514,200 by William Painter.
1894, January 1. P-114 Folding gay nineties legs. German patent 21,718 by Steinfeld & Reimer.
1894, July 10. P-70 Waiter's Friend. Patent 522,672 by Charles Puddefoot.
1896, June 9, and September 7, 1897. P-64 "Champion" bar mounted corkscrew. Design patent 25,607 and patent 589,574 Michael Redlinger.
1897, March 23. P-17 Wood handle corkscrew. Patent 579,200 by Edwin Walker.
1897, June 1. L-2 Mini bottle roundlet corkscrew. Patent 583,561 by William A. Williamson.
1898, September 20. P-87 Peg and worm corkscrew. Patent 611,046 by Edwin Walker.
1898, December 13. P-51 and P-116 Wood handle corkscrew. Design patent 29,798 by William A. Williamson.
1900, April 17. P-10 Wood handle corkscrew. Patent 647,775 by Edwin Walker.
1900, September 4. L-2 and M-25 bottle and bullet roundlet corkscrews. Patent 657,421 by Ralph W. Jorres.
1901, February 19. B-18, B-19, B-21, B-52, and B-63 openers. Design patent 34,096 by Augustus W. Stephens.
1901, May 28. P-111 "Shomee" bar mount corkscrew. Patent 675,032 by Albert Baumgarten
1904, July 19. P-68 "Sure Cut" can opener. Patent 765,450 by Frank White and Fred Winkler.
1905, June 27, and July 3, 1906. P-41 "Universal" folding corkscrew. Patents 793,318 and 824,807 by Harry W. Noyes.
1905, November 28. B-7 Handy Pocket Companion. Patent 805,486 by Julius T. Rosenheimer.
1906, August 14. M-52 Cap lifter with loop seal remover. Design patent 38,166 by John Hasselbring.
1906, December 25. P-23 "Yankee" can opener. Patent 839,229 by Charles G. Taylor.
1907, June 25. P-45 "Yankee No. 1" wall mount corkscrew. Patent 857,992 by Raymond B. Gilchrist.
1909, July 13. G-5 Cap lifter with flat handle. Patent 928,156 by Adolph Rydquist.
1909, October 12. N-7 and N-8 Cap lifter. Patent 936,678 by John L. Sommer.
1910, March 1. P-12 Wood handle corkscrew. Patent 950,509 by William R. Clough.
1911, September 11 M-1 and M-2 Lithographed cap lifter. Design patent 41,807 by Harry L. Beach.
1911, November 7. A-18 and A-42 Fish. Design patent 41,894 by John L. Sommer.
1911, November 7. A-13 and A-52 Automobile. Design patent 41,895 by John L. Sommer.
1912, March 12. A-28 and A-29 Bottle. Design patent 42,305 by John L. Sommer.
1912, March 12. A-7 and A-35. Fancy lady's boot. Design patent 42,306 by John L. Sommer.
1912, March 26. C-30 Cap lifter. Design patent 42,368 by Frank Mossberg.
1912, June 18. O-4 "Never Chip." Patent 1,029,645 by Harry L. Vaughan.
1912, October 6. B-70 Composite Tool. Patent 1,040,564 by A. W. Merrill.
1912, October 8, and October 7, 1913. N-49 Cap lifter, screwdriver, cigar cutter, Prest-O-Lite key, and watch fob. Patent by Arthur Merrill.
1912, November 26. F-6 "Four in 1 Handy Tool." Design patent 43,278 by Thomas Harding.
1913, April 8. P-83 "Yankee No. 7" bar mounted corkscrew. Patent 1,058,361 by R. B. Gilchrist.
1913, June 17. A-4 and A-5 Girl clothed (calendar) and nude (Early Morn). Design patent 44,226 by Harry L. Vaughan.
1913, November 25. A-30 and A-53 Dancer legs. Design patent 44,945 by Harry L. Vaughan.

1914, April 14. A-34 Powder Horn. Design patent 45,678 by John L. Sommer.
1914, August 18. A-9 Baseball player in pitching position. Design patent 46,298 by John L. Sommer.
1914, August 25. F-18 Cap lifter with ice pick. Design patent 46,311 by Thomas Harding.
1914, November 10. P-100 "Cap Remover." Patent 1,116,509 entitled "Cap Remover" by Josephine Spielbauer.
1914, November 24. F-13 Cap lifter and cake server. Design patent 46,702 by John L. Sommer and Thomas Harding.
1914, December 8. Type A-1 Bathing girl, mermaid or surf-girl. Design patent 46,762 by Harry L. Vaughan.
1915, January 12. B-28 Cap lifter with screw driver and cigar cutter. Patent 1,124,288 by Elvah V. Bulman.
1915, February 23. F-12 Cap lifter and slotted ladle. Design patent 47,016 by John L. Sommer.
1915, August 17. E-8 Single wire loop. Patent 1,150,083 by Edwin Walker.
1916, February 16. C-1 Cap lifter. Design patent 48,550 by Nelson Jacobus.
1916, December 5. B-13 Nifty. Patent 1,207,100 by Harry L. Vaughan.
1919, December 9. N-10 Cap lifter with bottle stopper. Patent 1,324,256 by William B. Langan.
1924, April 15. H-1, H-2, and H-3. Over-the-top cap lifter. Design patent 1,490,149 by Harry L. Vaughan.
1928, August 14. B-35 Pocket opener and corkscrew. Patent 1,680,291 by Thomas Harding.
1928, December 13, and February 12, 1929. H-4 Over-the-top cap lifter with folding corkscrew. Patents 1,695,098 and 1,701,950 by William Hiering.
1929, May 21. R-11 Figural parrot cap lifter and corkscrew. Design patent 78,544 by M. D. Avillar.
1929, May 7. O-8 Wall mounted cap lifter. Patent 1,711,678 by Thomas Harding.
1929, November 12. R-10 Figural dog cap lifter. Design patent 79,877 by M. D. Avillar.
1931, August 25. N-12 Cap lifter and lighter. Patent 1,820,131 issued for the lighter to Howard L. Fischer.
1931, September 22. F-2 Spoon with cap lifter. Design patent 85,178 by Thomas Harding.
1933, November 7. G-9 Cap lifter with wood handle. Patent 1,934,594 by Harry G. Edlund.
1934, February 27. C-22 Cap lifter. Design patent 91,635 by Ferdinand Neumer.
1935, April 2. I-7 The original Cap lifter/can piercer. Patent 1,996,550 by Dewitt F. Sampson and John M. Hothersall.
1935, May 21. I-25 Cap lifter/can piercer. Patent 2,002,173 by Bernard E. Dougherty.
1935, September 8. I-15 Cap lifter/can piercer. Patent 2,053,637 by Herbert Schrader.
1935, October 22. B-24 Vaughan's "Never Slip" bottle opener. Patent 2,018,083 issued to James A. Murdock.
1935, October 29. I-10 Cap lifter/can piercer. Patent 2,019,099 by Francis H. Schwartz.
1936, February 4. N-62 Cap lifter with bottle resealer. Design patent 98,486 by M. E. Trollen.
1940, January 11. J-7 Folding can piercer. Patent 2,188,352 by Dewitt F. Simpson and John M. Hothersall.
1940, April 16. N-45 Round knife with cutting blade and opener blade. Design patent 119,965 by Antonio Paolantonio.
1943, June 29. S-2 Cone top beer can with cap lifter. Patent 2,322,843 by Gerald Deane, New York.
1943, November 2. O-5 "Starr." Patent 2,033,088 by Raymond M. Brown.
1945, December 25. I-4 Cap lifter/can piercer. Design patent 143,327 by Michael J. LaForte.
1948, February 3. M-19 Cap lifter and muddler. Design patent 148,535 by Frank E. Hamilton.
1949, April 12. N-2 Cap lifter with bowling pin shape handle. Design patent 153,349 by Oscar Galter.
1949, August 6. K-1 Heavy cast iron opener. Design patent 154,792 by Harry L. Morris.
1949, September 20. J-9 Can piercer with tubular handle. Design patent 155,314 by Joseph G. Pessina.
1950, August 1, and September 4, 1951. I-22 "Easi-Ope" Cap lifter/can piercer. Design patent 164,448 and Mechanical patent 2,517,443 by Harland R. Ransom.
1950, October 17. O-9 Wall mounted cap lifter. Design patent 160,453 by Davis J. Ajouelo.
1950, October 31. F-3 Spoon with cap lifter and can piercer. Design patent 160,1950 issued to Le Emmette V. De Fee.
1950, December 26. A-39 Turtle with three screwdrivers. Design patent 161,321 by Le Emmette V. De Fee.
1951, May 1. Z-1 "Miracle Opener." Patent 2,651,511 by Joseph A. Talbot.
1952, October 28. I-20 Cap lifter/can piercer. Design patent 168,053 by Michael J. LaForte.
1953, December 1. M-39 "Tap Boy." Design patent 170,999 by Michael J. LaForte.
1956, January 3. N-26 Cap lifter, lid prier, and jar top opener. Design patent 176,518 by James L. Hvale.
1956, December 11. J-8 Can piercer/cap lifter. Patent 2,773,272 by George R. Harrah.
1957, January 1. N-3 "Derby Duke." Patent 2,779,098 by Edward J. Pocoski and William G. Hennessy.
1959, January 7. K-12 Can opener. Design patent 181,856 by Eric Johnson.
1959, March 18. S-1 Bottle with opener formed in glass in the bottom. Patent 2,992,574 by Werner Martinmaas.
1961, June 6. N-32 Cap lifter and can piercer on a fishing lure. Patent 2,986,812 by William Arter, Jr. and Robert J. Clouthier.
1974, April 16. Q-3 Plastic grip with metal cap lifter insert. Design patent 231,313 by G. John Heelan.
1977, December 6. N-40 Belt buckle with cap lifter. Design patent 246,552 by Daniel Baughman.

Cleaning Openers

Many collectors consider cleaning openers a sin and do not even attempt to clean their openers. For those who wish to clean openers here are several methods.

In the first issue of *Just for Openers* (April, 1979), Ed Kaye wrote: "Each opener (if not mint) I get from whatever source, is first wire brushed, then dipped in Naval Jelly, then rinsed in cold water, transferred to extremely hot water, hand dried, and finally buffed."

In July, 1984, Jack Lennon reported that he cleaned openers with fine steel wool and car polish compound. He would then clear spray them.

In July, 1986, Jim Hollinger's letter contained this explanation on how he cleaned openers: "I use various methods for cleaning my openers depending on the original condition. The various steps used in cleaning include:

1. Naval Jelly: For very poor condition openers. Openers such as this usually are not worth the effort due to excess pitting.
2. Fine Wire Brush: Fine wire brush is used in a drill press for removing spotty rust areas. A fine wire brush does not scratch the opener finish.
3. Polishing Compound: Automobile paint scratch and blemish remover is used with a soft cloth to buff all openers. This removes dirt and discoloration.
4. Paste Car Wax: All my openers receive a coat of car wax as the final step. This returns the original sheen of the plating as well as it leaves a protective coating against further corrosion."

In July, 1990, John Stanley presented this process:

1. Prepare a solution of oxalic acid (can collectors use this) using about 3 tablespoons of crystals for 2 pints of water. I also use a narrow lidded glass container so the opener will not lay flat in the bottom of the jar. The oxalic acid can be purchased at your local drugstore (they may have to special order) and the cost is about $6 for 16 ounces.
2. Depending upon the condition I will soak an opener in the solution from 30 minutes to 4 hours (for one with a lot of rust). I cleaned an A-13 Car for Gary Deachman at the convention soaking it for about 4 hours. It came out so nice he didn't want to trade the opener to me.
3. At the end of the soaking time, remove the opener from the solution using an old ice pick or awl (unless opener has no key ring hole then you need to pour acid into another container and remove opener). Careful not to allow skin to come in contact with acid. Thoroughly rinse skin if contacted.
4. I then scrub the opener with a soapy "SOS" pad. I dry the opener thoroughly with paper towels. If the opener has a moving part, I will add some machine oil.

Notes

1. Do not try to clean any openers with enameling using the acid solution.
2. If an opener is pitted, I use a Dremel (small electric drill). The Dremel is used by jewelers and anyone needing a small drill but it can leave brush marks. The wire brush attachment is very handy for cleaning pitted openers. You cannot remove the pit marks, but the discoloration can be cleaned off.
3. An alternative to the Dremel is using a regular drill with a "copper" wire brush.
4. Cleaning openers with "celluloid" parts is very tough. Use warm soapy water to clean the celluloid part. A soapy SOS or just a piece of fine steel wool is used to clean the metal opener. Celluloid is very tough, because, if the lettering is worn, you are just plain out of luck.
5. Enameled Openers present a real challenge because any rubbing of the enamel will take it right off.

Other suggestions

Harold Queen: "When using Oxalic Acid it is safe to use with your bare hands, since the acid is diluted with water. I use 2 heaping teaspoons to about 6 oz. of water." He then uses a soapy SOS pad and finally uses a fine wire brush to complete the process.

Joe Knapp: "I use Dow Bathroom Cleaner with Scrubbing Bubbles and a fine brass wire brush. I complete the process with chrome polish."

Ollie Hibbeler: "I use white distilled vinegar, soaking the opener for about 12-15 hours."

Resources

Books

Bull, Donald A. *Bull's Pocket Guide to Corkscrews*. Atlgen, Pennsylvania, USA: Schiffer Publishing Ltd., 1999.

Bull, Donald A. *The Ultimate Corkscrew Book*. Atlgen, Pennsylvania, USA: Schiffer Publishing Ltd., 1999.

Bull, Donald, and Manfred Friedrich. *The Register of United States Breweries 1876-1976, Volumes I & II*. Trumbull, Connecticut, USA: Bull, 1976.

Bull, Donald, Manfred Friedrich, and Robert Gottschalk. *American Breweries*. Trumbull, Connecticut, USA: Bullworks, 1984.

Bull, Donald. *A Price Guide to Beer Advertising Openers and Corkscrews*. Trumbull, Connecticut, USA: Bull, 1981.

Bull, Donald. *Beer Advertising Openers - A Pictorial Guide*. Trumbull, Connecticut, USA: Bull, 1978.

Goins, John. *Encyclopedia of Cutlery Markings*. Knoxville, Tennessee: Knife World Publications, 1986.

Kaye, Edward R., and Donald A. Bull. *The Handbook of Beer Advertising Openers and Corkscrews*. Sanibel Island, Florida: Kaye, 1994.

Levine, Bernard. *Levine's Guide to Knives and Their Values*. Iola, Wisconsin: Krause Publications, 1997.

O'Leary, Fred. *Corkscrews: 1000 Patented Ways to Open a Bottle*. Atlgen, Pennsylvania, USA: Schiffer Publishing Ltd., 1996.

Petretti, Allan. *Petretti's Soda Pop Collectibles Price Guide The Encyclopedia of Soda-Pop Collectibles*. Antique Trader Books, 1996.

Stanley, John R., Edward R. Kaye, and Donald A. Bull. *The 1998 Handbook of United States Beer Advertising Openers and Corkscrews*. Chapel Hill, North Carolina: John Stanley, 1998.

Van Wieren, Dale P., Donald Bull, Manfred Friedrich, and Robert Gottschalk. *American Breweries II*. West Point, Pennsylvania, USA: Eastern Coast Breweriana Association, 1995.

Newsletters

Bull, Donald. *Just for Openers, Issues 1-20*. Trumbull, Connecticut, USA: Bull, January, 1979 through October, 1983.

Kaye, Edward R. *Just for Openers, Issues 21-40*. Sanibel Island, Florida: Kaye, January, 1984 through October, 1988.

Santen, Art. *Just for Openers, Issues 41-60*. St. Louis, Missouri: Santen, January, 1985 through October, 1993.

Stanley, John R. *Just for Openers, Issues 61-80*. Chapel Hill, North Carolina: Stanley, January, 1994 through October, 1998.

Type Index

⋆ Deleted: Type already assigned type number or discovered to be foreign.